Lecture Notes in Computer Science   9495

Commenced Publication in 1973
Founding and Former Series Editors:
Gerhard Goos, Juris Hartmanis, and Jan van Leeuwen

Editorial Board

More information about this series at http://www.springer.com/series/7409

Jianfeng Zhan · Rui Han
Roberto V. Zicari (Eds.)

Big Data Benchmarks, Performance Optimization, and Emerging Hardware

6th Workshop, BPOE 2015
Kohala, HI, USA, August 31 – September 4, 2015
Revised Selected Papers

 Springer

Editors
Jianfeng Zhan
Institute of Computing
Chinese Academy of Sciences
Beijing
China

Roberto V. Zicari
Goethe Universität Frankfurt
Frankfurt
Germany

Rui Han
ICT, Chinese Academy of Sciences
Beijing
China

ISSN 0302-9743   ISSN 1611-3349 (electronic)
Lecture Notes in Computer Science
ISBN 978-3-319-29005-8   ISBN 978-3-319-29006-5   (eBook)
DOI 10.1007/978-3-319-29006-5

Library of Congress Control Number: 2015959926

LNCS Sublibrary: SL3 – Information Systems and Applications, incl. Internet/Web, and HCI

Printed on acid-free paper

This Springer imprint is published by SpringerNature
The registered company is Springer International Publishing AG Switzerland

Preface

In recent years, data explosion has become an inevitable trend as the world is more connected than ever before. In the era of big data, the complexity, diversity, frequently changed workloads, and rapid evolution of big data systems give rise to new challenges with traditional benchmarks being inadequate. As architecture, systems, and data management experts from both industry and academia communities pay greater attention to innovative systems and architectures, the pressure of benchmarking and evaluating big data systems increases. Benchmarking big data systems not only motivates system engineering efforts that optimize the performance and efficiency of system execution, but also promotes improvement in big data technology and stimulates investigations into new data management systems, hardware architectures, operating systems, and programming systems.

This book includes papers from the 6th Workshop on Big Data *B*enchmarks, *P*erformance *O*ptimization, and *E*merging Hardware (BPOE-6) (http://prof.ict.ac.cn/bpoe_6_vldb), which was co-located with VLDB 2015 (http://vldb.org/2015/), a premier conference on data management, database, and information systems. The workshop focuses on architecture and system support for big data systems, aiming at bringing researchers and practitioners from data management, architecture, and systems research communities together to discuss the research issues at the intersection of these areas. This book also invites three papers from several industry partners, including two papers describing tools used in system benchmarking and monitoring and one paper discussing principles and methodologies in existing big data benchmarks.

The call for papers for the BPOE-6 workshop attracted a number of high-quality international submissions. During a rigorous review process, in which each paper was assessed by at least four experts, we selected eight out of ten submissions for presentation at BPOE-6. In addition, several prestigious keynote speakers were invited, including Prof. Xiaodong Zhang, Ohio State University (http://web.cse.ohio-state.edu/~zhang//), whose topic was "Fast Data Accesses in Memory and Storage," and Prof. Dhabaleswar K. (DK) Panda, Ohio State University (http://www.cse.ohio-state.edu/~panda/), whose topic was "Accelerating and Benchmarking Big Data Processing on Modern HPC Clusters HPC and Big Data."

We are very grateful to the efforts of all authors related to writing, revising, and presenting their papers at BPOE workshops. Finally, we appreciate the indispensable support of the BPOE Program Committee and thank them for their efforts and contributions in maintaining the high standards of the BPOE workshop.

November 2015

Jianfeng Zhan
Rui Han
Roberto V. Zicari

Organization

Program Co-chairs

Jianfeng Zhan	ICT, Chinese Academy of Sciences and University of Chinese Academy of Sciences, China
Rui Han	ICT, Chinese Academy of Sciences, China
Roberto V. Zicari	Frankfurt Big Data Lab, Goethe University, Frankfurt, Germany

Steering Committee

Christos Kozyrakis	Stanford University, USA
Xiaofang Zhou	University of Queensland, Australia
Dhabaleswar K. (DK) Panda	Ohio State University, USA
Aoying Zhou	East China Normal University, China
Raghunath Nambiar	Cisco, USA
Lizy Kurian John	University of Texas at Austin, USA
Xiaoyong Du	Renmin University of China, China
Ippokratis Pandis	IBM Almaden Research Center, USA
Xueqi Cheng	ICT, Chinese Academy of Sciences, China
Bill Jia	Facebook, USA
Lidong Zhou	Microsoft Research Asia, China
H. Peter Hofstee	IBM Austin Research Laboratory, USA
Alexandros Labrinidis	University of Pittsburgh, USA
Cheng-Zhong Xu	Wayne State University, USA
Jianfeng Zhan	ICT, Chinese Academy of Sciences, China
Guang R. Gao	University of Delaware, USA
Yunquan Zhang	ICT, Chinese Academy of Sciences, China

Program Committee

Bingsheng He	Nanyang Technological University, Singapore
Xu Liu	College of William and Mary, USA
Rong Chen	Shanghai Jiao Tong University, China
Weijia Xu	Texas Advanced Computing Center, University of Texas at Austin, USA
Lijie Wen	School of Software, Tsinghua University, China
Xiaoyi Lu	The Ohio State University, USA
Yuqing Zhu	Institute of Computing Technology, Chinese Academy of Sciences, China
Yueguo Chen	Renmin University, China

Edwin Sha	Chongqing University, China
Mingyu Chen	Institute of Computing Technology, Chinese Academy of Sciences, China
Zhenyu Guo	MSRA, China
Tilmann Rabl	University of Toronto, Canada
Farhan Tauheed	EPFL, Switzerland
Chaitan Baru	San Diego Supercomputer Center, UC San Diego, USA
Seetharami Seelam	IBM, USA
Rene Mueller	IBM Almaden Research Center, USA
Cheqing Jin	East China Normal University, China

Contents

Emerging Hardware

Benchmarking

Revisiting Benchmarking Principles and Methodologies for Big Data Benchmarking

Liutao Zhao[1]([⊠]), Wanling Gao[2], and Yi Jin[1]

[1] Key Laboratory of Cloud Computing Technology and Application,
Beijing Computing Center, Beijing, China
{zhaolt,jinyi}@bcc.ac.cn
[2] State Key Laboratory of Computer Architecture,
Institute of Computing Technology, Chinese Academy of Sciences, Beijing, China
gaowanling@ict.ac.cn

Abstract. Benchmarking as yardsticks for system design and evaluation, has developed a long period and plays a pivotal role in many domains, such as database systems and high performance computing. Through prolonged and unremitting efforts, benchmarks on these domains have been reaching their maturity gradually. However, in terms of emerging scenarios of big data, its different properties in data volume, data types, data processing requirements and techniques, make that existing benchmarks are rarely appropriate for big data systems and further make us wonder how to define a good big data benchmark. In this paper, we revisit successful benchmarks in other domains from two perspectives: benchmarking principles which define fundamental rules, and methodologies which guide the benchmark constructions. Further, we conclude the benchmarking principle and methodology on big data benchmarking from a recent open-source effort – BigDataBench.

1 Introduction

Emerging big data and corresponding systems desperately need a big data benchmark suite, to evaluate and compare big data storage and processing systems. However, the properties of big data (Volume, Variety, Velocity) make big data benchmarking face great challenges. First, a design strategy of one-size-fits-all or one-size-fits-a-bunch. Second, a trade-off between benchmark coverage and cost. Third, great pressure on adapting rapid evolution of workloads, data sets and software stacks. How to define principles and methodologies for big data benchmarking are hotspot issues.

Benchmarking principles specify fundamental rules [4]. Jim Gray proposed four principles to make a good domain-specific benchmark [8], namely, relevant, portable, scaleable and simple. Likewise, revisiting successful benchmarks targeting at different domains or systems, such as SPEC benchmarks [2] for newest generation of high-performance computers, TPC benchmarks [3] for transaction processing and database systems, PARSEC benchmark [5] for multi-core and multiprocessor systems, implicit benchmarking principles exist. For example, to

J. Zhan et al. (Eds.): BPOE 2015, LNCS 9495, pp. 3–9, 2016.
DOI: 10.1007/978-3-319-29006-5_1

benchmark transaction processing and database systems, one primary principle for TPC benchmarks is relevant with the target domain.

Benchmarking methodologies provide design strategies and guidelines. A top-down approach and a bottom-up approach are two general methodologies to construct a benchmark [5]. Top-down means select representative programs and construct a sample of the application space, while bottom-up means recreate diverse range of characteristics of program behavior and solve the problem of difficulties to state representativeness.

In this paper, we revisit benchmarking principles and methodologies of successful benchmarks in traditional domains, i.e. SPEC benchmarks, TPC benchmarks, PARSEC benchmark, and further expand to big data scenarios, from a recent open-source big data benchmarking effort – BigDataBench [13].

2   Benchmarking Principles

In this section, we revisit the benchmarking principles of three successful benchmarks in detail, i.e. SPEC benchmarks, TPC benchmarks, and PARSEC benchmark, and further discuss the principles for big data benchmarking.

2.1   TPC Benchmarks

The Transaction Processing Performance Council (TPC) [3] has defined a series of transaction processing and database benchmarks following three considerations [9]: (1) No single metric can measure comprehensively, and one single metric not always suits for different systems [8]. (2) The more general the benchmark, the less useful it is for anything in particular; (3) A benchmark is a distillation of the essential attributes of a workload [9]. As a guide, TPC proposes seven desirable attributes for the development of benchmarks [9].

Relevant: Benchmark should be meaningful within the target domain.

Understandable: Benchmark should be simple and easy to understand.

Good metrics: Benchmark should define linear, orthogonal and monotonic metrics.

scaleable: Benchmark should be applicable to systems with different scale.

Coverage: Benchmark should not oversimplify the typical environment.

Acceptance: Benchmark should be accepted and recognized as useful by correlative vendors and users.

Portable: Benchmark should not be limited to one system or architecture.

2.2   SPEC Benchmarks

Systems Performance Evaluation Cooperative (SPEC) has proposed a set of benchmarks for high performance computers [2], such as SPEC CPU for measuring compute-intensive performance of different systems. SPEC follows five explicitly principles [4] to make it a well-known benchmark suite.

Application-oriented: Different from synthetic benchmarks, SPEC constructs benchmarks from end-user applications, to approach as nearly as possible the realities of real applications.

Portability: SPEC benchmarks generally use platform neutral programming languages to achieve portability to various systems and architectures.

Repeatable and reliable: SPEC specifies a set of run rules to make a wide range of results repeatable and reliable, such as rules must be adhered to for base optimizations.

Consistency and fairness: Likewise, SPEC should define strict rules about how to run and how to report results, to ensure the consistency and fairness of benchmarks.

2.3   PARSEC Benchmark

PARSEC is a benchmark suite composed of multithreaded programs [5]. The constructing of PARSEC benchmark follows five principles to achieve ease of use and flexibility.

Automatization: Benchmark should allow the user to perform standard tasks in an automated and centralized way, and let them get away from the details.

Modularity: Benchmark introduces modularity to simplify its handling and add new programs and inputs easily.

Abstraction: Benchmark should abstract from details of programs and allow users to conduct experiments without having to know much about the workloads except how to compile and execute.

Encapsulation: Benchmark should make it usable in an easily understandable way, by encapsulated the details of the inner workings of a workload in standardized configuration files.

Logging: Benchmark should allow the user to recreate their steps by automatically logging important information.

2.4   Big Data Benchmarking Community

The Big Data Benchmarking Community (BDBC) [1] organizes a group of individuals and organizations with the same interests on big data benchmarking. They have hold a series of workshops to facilitate discussion since 2012. The Fifth Workshop on Big Data Benchmarking (5th WBDB) which held in 2014, has proposed four principles for a successful big data benchmark.

Simple: The benchmark should be simple to implement and execute.

Cost effective: Not only benchmark construction but also benchmark execution should be cost-controlled.

Timely: The benchmark should upgrade its version timely to keep pace with the rapid evolution of big data domains.

Verifiable: The benchmark results must be reliable and verifiable through independent verification method.

Corresponding to big data benchmarking challenges mentioned above, these four principles are necessary, but not enough. Big data shows strikingly different properties in terms of data volume, data semantics and data types, thus big data benchmarking should involve diversity and complexity of big data.

2.5   BigDataBench

BigDataBench is an open-source big data benchmark suite [13], it focuses on broad application domains and diverse data models and workloads types. Big-DataBench defines the following five principles.

Data-centric: In big data era, diversified data models take the place of a single pattern, and new data-centric systems and architectures are designed to process and manage the big volume of data. Consequently, a good big data benchmark must be data-centric, not only including different types of real data, but also supporting the expansion of data scale.

Application-centric: Seltzer et al. [12] and Chen et al. [6] have illustrated the importance of using application-specific benchmarks to produce meaningful performance numbers in the context of real applications.

Coverage: Big data incarnates its wide coverage in ever-increasing application domains, data models, workloads and software stacks. Likewise, to evaluate big data systems and architectures, benchmarks should have a broad range of coverage.

Scaleable and Extensible: Big data benchmark should keep pace with the fast growth and rapid evolution of big data, which indicates its demand of scalability and extensibility. Scaleable means having a good adaptability of different cluster scales. Extensible means easy to add new workloads and support new software stacks.

Usability: Usability means easy to deploy, configure and run for users.

3   Benchmarking Methodologies

In this section, we revisit benchmarking methodologies of TPC-C [3] benchmark, and BigDataBench [13].

3.1   TPC-C Methodology

TPC-C is an on-line transaction processing (OLTP) benchmark with multiple transaction types, complex database and overall execution structure. The constructing of TPC-C introduces the concept of abstractions – functions of abstraction and functional workload model [6].

Functions of Abstraction. Through investigating multiple transactions in OLTP application domain, TPC-C concentrates on finding the compute units that frequently appeared with repetitions or similarities, such as duplicate operations and processing flow on different tables. After that, TPC-C finally abstracts five units of computation within an order-entry environment, with different proportion of read and write.

Functional Workload Model. Functional workload model includes functions of abstraction, data set and load pattern. Among which, data set involves data structure, inter-dependence and content. Load pattern involves execution frequency, distribution and arrival rate of each function of abstraction. That is to say, combining five primitive transactions, with specifying scheduling mode and defining data structure, a functional workload model can be formed.

3.2   Methodology of BigDataBench

Following the above five principles, BigDataBench [13] takes an incremental and iterative approach as illustrated in Fig. 1.

First, it investigates important or emerging application domains according to widely acceptable metrics. Second, it understands selected application domains from the perspectives of data sets and workloads [14], including data models of different types, i.e. structured, un-/semi-structured data sets and semantics, i.e. text, table, graph, image and workload dwarfs of units of computation that frequently appear in big data analytics [7].

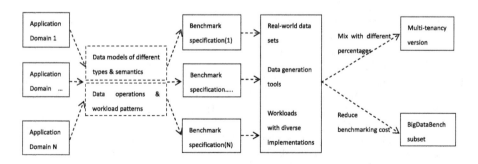

Fig. 1. Methodology of BigDataBench.

Third, it defines benchmark specifications [10] for each application domain to guide its implementations from aspects of data model description and important workloads extraction. Fourth, three aspects are considered to implement BigDataBench: representative and diverse real-world data sets, big data generator suite which preserve characteristics of real-world data [11], and important workloads implementation with competitive techniques.

Finally, it provides two versions for specific requirements and purposes: multi-tenancy version with different percentages of service and analytics workloads, and BigDataBench subset with a small number of representative workloads according to workload characteristics from a specific perspective.

4   Conclusion

Big data benchmarking is important and challenging. Revisiting successful benchmarks in other domains, we find that most of them specify benchmark principles according to domain-oriented benchmarking requirements, and define methodologies accord with principles. In terms of big data domain, the benchmarking principles should take its complexity and diversity into consideration, moreover, the methodologies should be capable to keep pace with the rapid evolution of big data domains.

Acknowledgements. This work is supported by the National High Technology Research and Development Program of China (Grant No. 2015AA015308), the Major Program of National Natural Science Foundation of China (Grant No. 61432006), and the Key Technology Research and Development Programs of Guangdong Province, China (Grant No. 2015B010108006).

References

1. Bdbc. http://clds.sdsc.edu/bdbc
2. Spec. https://www.spec.org/
3. Tpc. http://www.tpc.org
4. Angles, R.: Benchmark principles and methods. In: Linked Data Benchmark Council (LDBC). Project No 317548, European Community's Seventh Framework Programme FP7 (2012–2014)
5. Bienia, C., Li, K.: Benchmarking Modern Multiprocessors. Princeton University, New York (2011)
6. Chen, Y., Raab, F., Katz, R.: From TPC-C to big data benchmarks: a functional workload model. In: Rabl, T., Poess, M., Baru, C., Jacobsen, H.-A. (eds.) WBDB 2012. LNCS, vol. 8163, pp. 28–43. Springer, Heidelberg (2014)
7. Gao, W., Luo, C., Zhan, J., Ye, H., He, X., Wang, L., Zhu, Y., Tian, X.: Identifying dwarfs workloads in big data analytics (2015). arXiv preprint arXiv:1505.06872
8. Gray, J.: Benchmark Handbook: For Database and Transaction Processing Systems. Morgan Kaufmann Publishers Inc., San Francisco (1992)
9. Levine, C.: TPC benchmarks. In: SIGMOD International Conference on Managementof Data - Industrial Session (1997)

10. Luo, C., Gao, W., Jia, Z., Han, R., Li, J., Lin, X., Wang, L., Zhu, Y., Zhan, J.: Handbook of bigdatabench (version 3.1) - a big data benchmark suite
11. Ming, Z., Luo, C., Gao, W., Han, R., Yang, Q., Wang, L., Zhan, J.: BDGS: A scalable big data generator suite in big data benchmarking. In: Rabl, T., Raghunath, N., Poess, M., Bhandarkar, M., Jacobsen, H.A., Baru, C. (eds.) Advancing Big Data Benchmarks. LNCS, vol. 8585. Springer, Heidelberg (2014)
12. Seltzer, M., Krinsky, D., Smith, K., Zhang, X.: The case for application-specific benchmarking. In: Proceedings of the Seventh Workshop on Hot Topics in Operating Systems, pp. 102–107. IEEE (1999)
13. Wang, L., Zhan, J., Luo, C., Zhu, Y., Yang, Q., He, Y., Gao, W., Jia, Z., Shi, Y., Zhang, S., et al.: BigDataBench: A big data benchmark suite from internet services. In: 2014 IEEE 20th International Symposium on High Performance Computer Architecture (HPCA), pp. 488–499. IEEE (2014)
14. Zhu, Y., Zhan, J., Weng, C., Nambiar, R., Zhang, J., Chen, X., Wang, L.: BigOP: generating comprehensive big data workloads as a benchmarking framework. In: Bhowmick, S.S., Dyreson, C.E., Jensen, C.S., Lee, M.L., Muliantara, A., Thalheim, B. (eds.) DASFAA 2014, Part II. LNCS, vol. 8422, pp. 483–492. Springer, Heidelberg (2014)

BigDataBench-MT: A Benchmark Tool for Generating Realistic Mixed Data Center Workloads

Rui Han[1](✉), Shulin Zhan[3], Chenrong Shao[4], Junwei Wang[5],
Lizy K. John[6], Jiangtao Xu[1], Gang Lu[1,2], and Lei Wang[1]

[1] Institute of Computing Technology, Chinese Academy of Sciences, Beijing, China
{hanrui,xujiangtao,lugang,wl}@ict.ac.cn
[2] University of Chinese Academy of Sciences, Beijing, China
[3] ICarsclub, Beijing, China
zhanshulin@ppzuche.com
[4] Xi'an Jiaotong University, Xi'an, China
scr1994@stu.xjtu.edu.cn
[5] Kingsoft Cloud, Beijing, China
wangjunwei@kingsoft.com
[6] Department of Electrical and Computer Engineering,
The University of Texas, Austin, TX, USA
ljohn@ece.utexas.edu

Abstract. Long-running service workloads (e.g. web search engine) and short-term data analysis workloads (e.g. Hadoop MapReduce jobs) colocate in today's data centers. Developing realistic benchmarks to reflect such practical scenario of mixed workload is a key problem to produce trustworthy results when evaluating and comparing data center systems. This requires using actual workloads as well as guaranteeing their submissions to follow patterns hidden in real-world traces. However, existing benchmarks either generate actual workloads based on probability models, or replay real-world workload traces using basic I/O operations. To fill this gap, we propose a benchmark tool that is a first step towards generating a mix of actual service and data analysis workloads on the basis of real workload traces. Our tool includes a combiner that enables the replaying of actual workloads according to the workload traces, and a multi-tenant generator that flexibly scales the workloads up and down according to users' requirements. Based on this, our demo illustrates the workload customization and generation process using a visual interface. The proposed tool, called BigDataBench-MT, is a multi-tenant version of our comprehensive benchmark suite BigDataBench and it is publicly available from http://prof.ict.ac.cn/BigDataBench/multi-tenancyversion/.

Keywords: Data center · Benchmark · Workload trace · Mixed workloads

© Springer International Publishing Switzerland 2016
J. Zhan et al. (Eds.): BPOE 2015, LNCS 9495, pp. 10–21, 2016.
DOI: 10.1007/978-3-319-29006-5_2

1   Introduction

In modern cloud data centers, a large number of tenants are consolidated to share a common computing infrastructure and execute a diverse mix of workloads. Benchmarking and understanding these workloads is a key problem for system designers, programmers and researchers to optimize the performance and energy efficiency of data center systems and to promote the development of data center technology. This work focuses on two classes of popular data center workloads [46]:

- *Long-running services.* These workloads offer online services such as web search engines and e-commerce sites to end users and the services usually keep running for months and years. The *tenants* of such workloads are *service end users.*
- *Short-term data analysis jobs.* These workloads process input data of many scales using relatively short periods (e.g. in Google and Facebook data centers, a majority (over 90 %) of analytic jobs complete within a few minutes [30, 46]). The *tenants* of such workloads are *job submitters.*

As data analysis systems such as Hadoop and Spark mature, both types of workloads widely co-locate in today's data centers, hence the pressure to benchmark and understand these mixed workloads rises. Within this context, we believe that it will be of interest to the data management community and a large user base to generate realistic workloads such that trustworthy benchmarking reflecting the practical data center scenarios can be conducted. Considering the heterogeneity and dynamic nature of data center workloads and their aggregated resource demands and arrival patterns, this requires overcoming two major challenges.

Benchmarking Using Actual Workloads Based on Real-World Workload Rraces. Data analysis jobs usually have various computation semantics (i.e. implementation logics or source codes) and input data sizes (e.g. ranging from KB to PB), and their behaviors also heavily rely on the underlying software stacks (such as Hadoop or MPI). Hence it is difficult to emulate the behaviors of such highly diverse workloads just using synthetic workloads such as I/O operations. On the other hand, generating workloads whose arrival patterns follow real-world traces is an equally important aspect of realistic workloads. This is because these traces are the most realistic data sources including both explicit and implicit arrival patterns (e.g. sequences of time stamped requests or jobs).

Benchmarking Using Scalable Workloads with Realistic Mixes. A good benchmark needs to flexibly adjust the scale of workloads to meet the requirements of different benchmarking scenarios. Based on our experience, we noticed that in many cases, obtaining real workload traces is difficult due to confidential issues. The limited trace data also restrict the scalability of benchmark. It is therefore challenging to produce workloads at different scales while still guaranteeing their realistic mix corresponding to real-world scenarios.

In this paper, we propose a benchmark tool that is a first step towards generating realistic mixed data center workloads. This tool, called BigDataBench-MT, is a multi-tenancy version of our open-source project BigDataBench, which is a comprehensive benchmark suite including 14 real-world data sets and 33 actual workloads covering five application domains [8,51]. The goal of BigDataBench-MT is not only supporting the generation of service and data analysis workloads based on real workload traces, but also providing a multi-tenant framework to enable the scaling up and down of such workloads with guarantee of their realistic mixes. Considering our community may feel interest in using these workloads to evaluate new system designs and implementations, our tool and the corresponding workload traces are publicly available from http://prof.ict.ac.cn/BigDataBench/multi-tenancyversion/.

2 Related Work

We now review existing data center benchmarks from three perspectives, as shown in Table 1.

Evaluated Platform. First of all, we classify data center benchmarks according to their targeted systems. We consider three popular camps of systems in today's data centers: (1) *Hadoop-related systems*: the great prosperity of the Hadoop-centric systems in industry brings a wide diversity of systems (e.g. Spark [6], HBase [1], Hive [2] and Impala [3]) on top of Hadoop MapReduce and HDFS [11] as well as a wide range of benchmarks specifically designed for these systems. (2) *Data stores*: parallel DBMSs (e.g. MySQL [12] and Oracle [15]) and NoSQL data stores (e.g. Amazon Dynamo [33], Cassandra [41] and Linkedin Voldemort [50]) also widely exist in data centers. (3) *Web services*: long-running web services such as Nutch search engine [5] and multi-tier cloud applications [36] are another important type of data center applications. These services usually have stringent response time requirement [37] and their request processing is distributed into a large number of service components for parallel processing, thus the service latency is determined by the tail latency of these components [32,38].

Workload Implementation Logic. Consider the complexity and diversity of workload behaviors in current data center systems, the implementation logic of existing data center benchmarks can be classified into three categories. The first category of benchmarks implement their workloads with algorithms. For example, HiBench [39] include workloads implemented with machine learning algorithms in Mahout [4]. The second category of benchmarks implement workloads using database operations such as reading, loading, joining, grouping, unifying, ordering, aggregating and spliting data. The third category of benchmarks implement workloads as I/O operations. For example, NNBench [13] and TestDFSIO [23] emulate I/O operations on Hadoop HDFS; GridMix [10] provides two workloads: LoadJob that performs I/O operations and SleepJob that sleeps the jobs; and SWIM [30] provides four workloads that stimulate the operations of Hadoop jobs to read, write, shuffle and sort data. We view the first two categories of workloads as *actual workloads*, because these workloads have semantics and they consume resources of processors, memories, caches and I/O bandwidths in execution. By contrast, workloads belonging to third category only consume I/O resources.

Workload Mix. Finally, we classify data center benchmarks into three categories from the perspective of workload mix. The first type of data center benchmarks either generate single workloads (e.g. WordCount [26], Grep [9] and Sort [19]) or generate multiple workloads individually (e.g. CALDA [44], AMPLab benchmark [7] and CloudSuite [34]). That is, these benchmarks donot consider workload mix. The second category of benchmarks generate synthetic mixes of workloads. Many benchmarks (e.g. PigMix [17], HcBench [47] and Big-Bench [35]) generate mixes of workloads by manually determining their proportions. Similarly, TPC benchmarks [24] design a query set as a synthetic mix of queries with different proportions. YCSB [31] uses a package to include a set of related workloads. MRBS decides the frequencies of different workloads using probability distributions such as a random distribution. Finally, third category of benchmarks generate a realistic mix of synthetic workloads whose arrival patterns faithfully follow real-world traces. For example, GridMix [10] and SWIM [21,30] first build a job trace to describe the realistic job mix by mining production loads, and then run synthetic I/O operations according to the trace. However, how to generate actual workloads on the basis of real workload traces is still an open question.

Table 1. Overview of data center benchmarks

Workload mix	Workload implementation logic		
	Algorithms	Database operations	I/O operations
No mix	WordCount[1] [26], Grep[1] [9], Sort[1] [19], Terasort[1] [22], HiBench[1] [39], TPCx-HS[1] [43], Graphalytics[1] [29], CloudSuite[4] [34]	MRBench[1][40], CALDA[2][44], AMPLab benchmark[2] [7], YCSB[2] [31], BG benchmark[2][28], CloudSuite[4] [34]	NNBench[1][13], TestDFSIO[1][23], HiBD[1][49]
Synthetic mix	HcBench[1][47], MRBS[1][48]	PigMix[1][17], HcBench[1][47], MRBS[1][48], BigBench[2][35], LinkBench[2][27], TPC benchmarks[2][24], TPC-W[3][25], BigDataBench[4][51]	HiBench[1][39], SPECWeb99[3][20]
Realistic mix			Gridmix[1][10], SWIM[1][21,30]

[1] Hadoop-related systems
[2] Data stores
[3] Web services
[4] All three types of systems

3   System Overview

Figure 1 shows the framework of our benchmark tool. It consists of three main modules. In the *Benchmark User Portal*, users can first specify their benchmarking requirements, including the machine type and number to be tested, and the types of workload to use. A set of workload traces following these requirements are then selected. The next step of *Combiner of Workloads and Traces* is to match the real workload and the selected workload traces, and outputs workload replaying scripts to guide the workload generation. Finally, the *Multi-tenant Workload Generator* extracts the tenant information from the scripts and constructs a multi-tenant framework to generate a mix of service and data analysis workloads.

In BigDataBench-MT, we employ the Sogou user query logs [18] as the basis to generate the service workload (i.e. the Nutch search engine [5]) and the Google cluster workload trace as the basis to generate data analysis workloads (i.e. Hadoop and Shark workloads). The Sogou trace records logs from 50 days and it includes over 9 million users and 43 million queries. The Google trace records logs from 29 days and 12,492 machines and it includes over 5 K users, 40 K workload types, 1000 K jobs and 144 million tasks. As a preprocessing step, we converted both traces into Impala databases (full version) and MySQL database (24-hour version) to facilitate the customization of benchmarking scenarios. In the following subsections, we describe the last two modules of our tool.

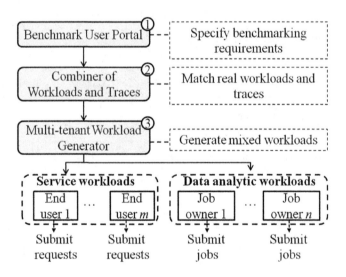

Fig. 1. The BigDataBench-MT framework

3.1   Combiner of Workload and Traces

The goal of the combiner is to extract the request/job arrival patterns from real-world traces and combine them with actual workloads. The combiner applies differentiated combination techniques to the service and data analysis workloads because their workload generations have different features.

Service Workloads. The generation of a service workload is determined by three factors: the request submitting time, the sequence of requests and the content of each request query. Take the web search engine for example, the combiner implements a request submitting service that automatically derives these factors from the Sogou trace and uses them to determine the request submission process.

Data Analysis Workloads. The generation of a data analysis workload is determined by four factors: the job submitting time, the workload type (i.e. the computation semantics and the software stack) and the input data (i.e. the data source and size). The current workload traces usually show the information of job submitting time but only provide *anonymous jobs* whose workload types and/or input data are unknown. Hence the basic idea of the combiner is to derive the workload characteristics of both actual and anonymous jobs and then match jobs whose workload characteristics are sufficiently similar. Table 2 lists the metrics used to represent workload characteristics, which reflect both jobs' performance (execution time and resource usage) and micro-architectural behaviors (CPI and MAI).

Table 2. Metrics to represent workload characteristics of data analysis jobs

Metric	Description
Execution time	Measured in seconds
CPU usage	Total CPU time per second
Total memory size	Measured in GB
CPI	Cycles per instruction
MAI	The number of memory accesses per instruction

Figure 2 shows the process of matching actual data analysis jobs and traces' anonymous jobs and it consists of two parallel sub-processes. First, the actual jobs with different input data sizes are tested and their metrics of workload characteristics are collected. In BigDataBench-MT, we provide auto-running scripts to collect performance metrics and hardware performance counters (Perf [16] and Oprofile [14] for Linux 2.6+ based systems) to obtain micro-architectural metrics. Using the testing results as samples, the combiner trains the multivariate regression model to describe the relationship between an actual job (including both its workload type and input size as the independent variables) and its workload characteristic metrics (one metric is a dependent variable). Second, the combiner views each anonymous job as an entity and the five workload characteristic metrics as its attributes, and employs the Bayesian Information Criterion (BIC)-based k-means clustering algorithm [45] to group anonymous jobs in the trace into different clusters.

Based on the constructed regression models and clusters, the combiner further matches each cluster to one actual job with a specific input data. In the matching, the coefficient of variation (CV) measure, defined as the ratio of the standard deviation σ to the mean μ, is used to describe the dispersion of jobs in the same cluster. The combiner iteratively tests actual jobs of different workload types and input sizes, and matches an actual job with a cluster under two conditions: (i) the CV of the cluster is smaller than a specified threshold (e.g. 0.5), which indicates the anonymous jobs in this cluster are closely similar to each other; (ii) the change in this CV is smaller than a threshold (e.g. 0.1) after the actual job is added to the cluster. This means the workload characteristics of the added job are sufficiently similar to those of the anonymous jobs in the cluster. If multiple matched actual jobs are found for one cluster, the combiner selects the job resulting the smallest CV change. Finally, the combiner produces workload replaying scripts as the output.

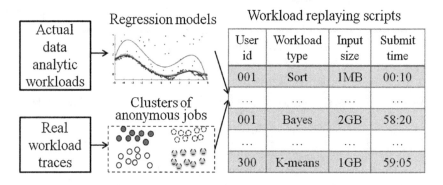

Fig. 2. The matching process of real and synthetic data analysis jobs

Note that BigDataBench-MT provides two ways of using the above combiner. First, it directly provides some workload replaying scripts, which are the combination results of representative actual workloads (e.g. Hadoop Sort and WordCount) and the Google workload trace. Second, it also supports benchmark users to directly use the above combination technique to match their own data analysis jobs with Google anonymous jobs.

3.2   Multi-tenant Workload Generator

Based on workload replaying scripts, the workload generator applies a multi-tenant mechanism to generate a mix of workloads using two steps. First, the generator extracts the tenant information from the scripts. For the service and data analysis workloads, this tenant information represents the number of concurrent end users and submitters of analytic jobs, respectively. Second, the generator creates a client for each tenant and emulates the scenarios that a number of end users/job submitters concurrently submit requests/jobs to the system. This multi-tenant framework allows the flexible adjustment of workload scales with guarantee of their realistic mixes. For example, benchmark users can double or halve the size of concurrent tenants, after which the distributions of requests/jobs submitted by these tenants still correspond to those in real workload traces.

4   Demonstration Description

4.1   Chosen Workloads and Workload Traces

In our demonstration, benchmark users want to evaluate their data center systems using a mix of service and data analysis workloads. The Nutch web search engine [5] is used as the example service workload and four Hadoop workloads are used as the example data analysis workloads. The chosen Hadoop workloads have a variety of workload characteristics: WordCount and Naïve Bayes classification are typical CPU-intensive workloads with integer and float point calculations; Sort is the typical I/O-intensive workload and PageIndex is the workload having similar demands for CPU and I/O resources. Both the data generators [42] and workloads in the demo can be obtained from BigDataBench [8].

Our demo uses a 24-h user query logs from Sogou, which include 1,724,264 queries from 519,876 end users, as the basis to generate realistic search engine service; and uses a 24-h cluster workload trace from Google, which includes 37,842 anonymous jobs from 2,261 job submitters, as the basis to generate realistic Hadoop jobs.

4.2   System Demonstration

BigDataBench-MT provides a visual interface in the *Benchmark User Portal* to help benchmark users make appropriate benchmarking decisions. This portal provides users necessary information, allows them input benchmarking requirements and executes system evaluations on their behalf. The whole process consists of three steps, as shown in Figs. 3, 4 and 5, respectively.

Step 1. Specification of tested machines and workloads. The first step of the demo presents an overview of workload traces (i.e. Sogou and Google traces) and the data center status, including the six types of machines, their machine number and configurations, and the user, job and task statistics in these machines. This information assists benchmark users to select the type and number of machines to be evaluated, and the workloads they want to use. Suppose users select *Type Four* of the machines with 2 process cores and 4 GB memory and 100 machines to be tested, the workload traces belonging to these machines are extracted and forwarded to the next step.

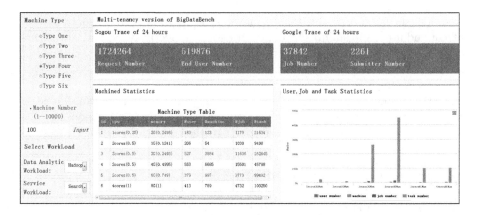

Fig. 3. System demonstration screenshots: step 1

Step 2. Selection of benchmarking period and scale. At this step, users have the option to select the period and scale for their benchmarking scenarios. To facilitate this selection, BigDataBench-MT shows the statistic information of both the service workload (including its number of requests and end users per second) and the data analysis workloads (including their number of jobs and average CPU, memory resource usages) at each of the 24 hs. Suppose users select the benchmarking period of 12:00 to 13:00 and the scale factor is 1 (that is, no scaling is needed). The workload traces belonging to this period are selected for step 3.

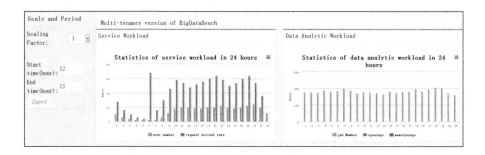

Fig. 4. System demonstration screenshots: Step 2

Step 3. Generation of mixed workloads. After both workloads and traces have been selected, the final step employs the combiner described in Sect. 3.1 to generate workload replaying scripts for both the service and data analysis workloads, and sends these scripts as feedback to users. In the matching of actual Hadoop jobs with anonymous ones, we tested each Hadoop workload type using 20 different input sizes to build the regression models. Based on the replaying scripts,

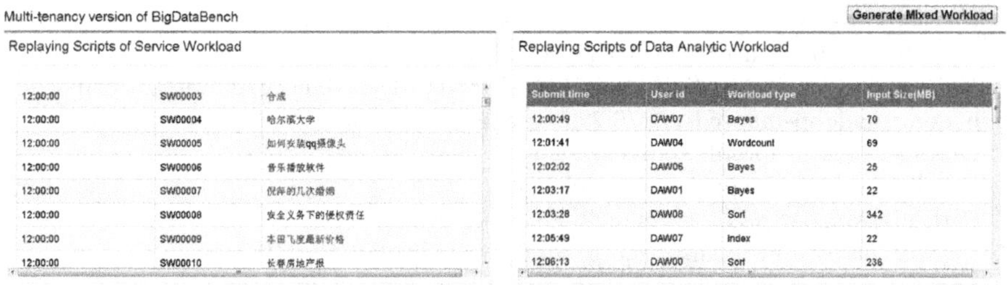

Multi-tenancy version of BigDataBench					Generate Mixed Workload		

Replaying Scripts of Service Workload			Replaying Scripts of Data Analytic Workload			

				Submit time	User id	Workload type	Input Size(MB)
12:00:00	SW00003	合成		12:00:49	DAW07	Bayes	70
12:00:00	SW00004	哈尔滨大学		12:01:41	DAW04	Wordcount	69
12:00:00	SW00005	如何安装qq摄像头		12:02:02	DAW06	Bayes	25
12:00:00	SW00006	音乐播放软件		12:03:17	DAW01	Bayes	22
12:00:00	SW00007	侵abc的几次缩indeed		12:03:28	DAW08	Sort	342
12:00:00	SW00008	安全义务下的侵权责任		12:05:49	DAW07	Index	22
12:00:00	SW00009	本田飞度最新价格		12:06:13	DAW00	Sort	236
12:00:00	SW00010	长春房地产网					

Fig. 5. System demonstration screenshots: step 3

benchmark users can press the "Generate mixed workload" button to trigger the multi-tenant workload generator, in which each tenant is an independent workload generator and multiple tenants generate a mix of realistic workloads.

5 Future Work

There are multiple avenues for extending the functionality of our benchmark tool. A first step will be to support more actual workloads. Given that there are 33 actual workloads in the BigDataBench and many workloads (e.g. Bayes classification and WordCount) have three versions of implementations (Hadoop, Spark and MPI), adding more workloads to BigDataBench-MT will be helpful to support wider benchmarking scenarios. We also plan to extend our multi-tenant workload generator to support different classes of tenants and allow users to apply different priority disciplines in workload generation.

Acknowledgements. This work is supported by the National High Technology Research and Development Program of China (Grant No. 2015AA015308), the National Natural Science Foundation of China (Grant No. 61502451), and the Key Technology Research and Development Programs of Guangdong Province, China (Grant No. 2015B010108006).

References

1. Apache hbase. http://hbase.apache.org/
2. Apache hive. https://cwiki.apache.org/confluence/display/Hive/Home
3. Apache impala. http://impala.io/
4. Apache Mahout. http://mahout.apache.org/
5. Apache Nutch. http://nutch.apache.org/
6. Apache spark. https://spark.apache.org/
7. Big data benchmark by amplab of UC berkeley. https://amplab.cs.berkeley.edu/benchmark/
8. BigDataBench. http://prof.ict.ac.cn/BigDataBench/
9. Grep. http://wiki.apache.org/hadoop/Grep
10. Gridmix. https://hadoop.apache.org/docs/r1.2.1/gridmix.html

11. Hadoop ecosystems. https://hadoopecosystemtable.github.io/
12. MySQL database. https://www.mysql.com/
13. Nnbench. . http://grepcode.com/file/repo1.maven.org/maven2/org.apache. hadoop/hadoop-mapred-test/0.22.0/org/apache/hadoop/hdfs/NNBench.java/
14. Oprofile. http://oprofile.sourceforge.net/
15. Oracle database. http://www.oracle.com/
16. Perf. https://perf.wiki.kernel.org/
17. PigMix. https://cwiki.apache.org/confluence/display/PIG/PigMix
18. Sogou user query logs. http://www.sogou.com/labs/dl/q-e.html
19. Sort. http://wiki.apache.org/hadoop/Sort
20. Specweb99 benchmark. https://www.spec.org/web2009/
21. Swim. https://github.com/SWIMProjectUCB/SWIM/wiki
22. Terasort. https://hadoop.apache.org/docs/current/api/org/apache/hadoop/ examples/terasort/package-summary.html
23. Testdfsio. https://support.pivotal.io/hc/en-us/articles/200864057-Running-DFSIO-mapreduce-benchmark-test/
24. TPC benchmarks. http://www.tpc.org/
25. TPC-W benchmark. http://www.tpc.org/tpcw/
26. WordCount. http://wiki.apache.org/hadoop/WordCount
27. Armstrong, T.G., Ponnekanti, V., Borthakur, D., Callaghan, M., Linkbench: a database benchmark based on the facebook social graph. In: SIGMOD 2013, pp. 1185–1196. ACM (2013)
28. Barahmand, S., Ghandeharizadeh, S., BG: A benchmark to evaluate interactive social networking actions. In: CIDR 2013. Citeseer (2013)
29. Capotă, M., Hegeman, T., Iosup, A., Prat-Pérez, A., Erling, O., Boncz, P., Graphalytics: A big data benchmark for graph-processing platforms. In: Proceedings of the GRADES 2015, pp. 7. ACM (2015)
30. Chen, Y., Alspaugh, S., Katz, R.: Interactive analytical processing in big data systems: A cross-industry study of mapreduce workloads. VLDB 5(12), 1802–1813 (2012)
31. Cooper, B.F., Silberstein, A., Tam, E., Ramakrishnan, R., Sears, R.: Benchmarking cloud serving systems with YCSB. In: SoCC 2010, pp. 143–154. ACM (2010)
32. Dean, J., Barroso, L.A.: The tail at scale. Commun. ACM 56(2), 74–80 (2013)
33. DeCandia, G., Hastorun, D., Jampani, M., Kakulapati, G., Lakshman, A., Pilchin, A., Sivasubramanian, S., Vosshall, P., Vogels, W., Dynamo: amazon's highly available key-value store. In: ACM SIGOPS Operating Systems Review, vol. 41, pp. 205–220. ACM (2007)
34. Ferdman, M., Adileh, A., Kocberber, O., Volos, S., Alisafaee, M., Jevdjic, D., Kaynak, C., Popescu, A.D., Ailamaki, A., Falsafi, B.: Clearing the clouds: A study of emerging workloads on modern hardware. Technical report (2011)
35. Ghazal, A., Rabl, T., Hu, M., Raab, F., Poess, M., Crolotte, A., Jacobsen, H.-A.: Bigbench: Towards an industry standard benchmark for big data analytics. In: SIGMOD 2013, pp. 1197–1208. ACM (2013)
36. Han, R., Ghanem, M.M., Guo, L., Guo, Y., Osmond, M.: Enabling cost-aware and adaptive elasticity of multi-tier cloud applications. Future Gener. Comput. Syst. 32, 82–98 (2014)
37. Han, R., Wang, J., Ge, F., Vazquez-Poletti, J.L., Zhan, J.: SARP: producing approximate results with small correctness losses for cloud interactive services. In: CF 2015, pp. 22. ACM (2015)

38. Han, R., Wang, J., Huang, S., Shao, C., Zhan, S., Zhan, J., Vazquez-Poletti, J.L., SARP: producing approximate results with small correctness losses for cloud interactive services. In: ICPP 2015, pp. 490–499. IEEE (2015)

39. Huang, S., Huang, J., Dai, J., Xie, T., Huang, B., The hibench benchmark suite: Characterization of the mapreduce-based data analysis. In: IEEE 26th International Conference on Data Engineering Workshops (ICDEW 2013), pp. 41–51. IEEE (2010)

40. Kim, K., Jeon, K., Han, H., Kim, S.-G., Jung, H., Yeom, H.Y.: Mrbench: A. benchmark for mapreduce framework. In: ICPADS 2008, pp. 11–18. IEEE (2008)

41. Lakshman, A., Malik, P.: Cassandra: a decentralized structured storage system. ACM SIGOPS Oper. Syst. Rev. **44**(2), 35–40 (2010)

42. Ming, Z., Luo, C., Gao, W., Han, R., Yang, Q., Wang, L., Zhan, J.: BDGS: A scalable big data generator suite in big data benchmarking. In: Rabl, T., Raghunath, N., Poess, M., Bhandarkar, M., Jacobsen, H.A., Baru, C. (eds.) Advancing Big Data Benchmarks. LNCS, vol. 8585, pp. 138–154. Springer, Heidelberg (2014)

43. Nambiar, R.: A standard for benchmarking big data systems. In: BigData 2014, pp. 18–20. IEEE (2014)

44. Pavlo, A., Paulson, E., Rasin, A., Abadi, D.J., DeWitt, D.J., Madden, S., Stonebraker, M.: A comparison of approaches to large-scale data analysis. In: SIGMOD 2009, pp. 165–178. ACM (2009)

45. Pelleg, D., Moore, A.W., X-means, et al.: Extending k-means with efficient estimation of the number of clusters. In: ICML 2000, pp. 727–734 (2000)

46. Reiss, C., Tumanov, A., Ganger, G.R., Katz, R.H., Kozuch, M.A.: Heterogeneity, dynamicity of clouds at scale: Google trace analysis. In: Proceedings of the Third ACM Symposium on Cloud Computing, pp. 7. ACM (2012)

47. Saletore, V., Krishnan, K., Viswanathan, V., Tolentino, M.E., Hcbench, et al.: Methodology, development, and characterization of a customer usage representative big data/hadoop benchmark. In: IISWC 2013, pp. 77–86. IEEE (2013)

48. Sangroya, A., Serrano, D., Bouchenak, S.: MRBS: towards dependability benchmarking for hadoop mapreduce. In: Caragiannis, I., Alexander, M., Badia, R.M., Cannataro, M., Costan, A., Danelutto, M., Desprez, F., Krammer, B., Sahuquillo, J., Scott, S.L., Weidendorfer, J. (eds.) Euro-Par Workshops 2012. LNCS, vol. 7640, pp. 3–12. Springer, Heidelberg (2013)

49. Shankar, D., Lu, X., Wasi-ur-Rahman, M., Islam, N., Panda, D.K.D.K.: A Micro-benchmark suite for evaluating hadoop mapreduce on high-performance networks. In: Zhan, J., Rui, H., Weng, C. (eds.) BPOE 2014. LNCS, vol. 8807, pp. 19–33. Springer, Heidelberg (2014)

50. Sumbaly, R., Kreps, J., Gao, L., Feinberg, A., Soman, C., Shah, S.: Serving large-scale batch computed data with project voldemort. In: FAST 2012, pp. 18–18. USENIX Association (2012)

51. Wang, L., Zhan, J., Luo, C., Zhu, Y., Yang, Q., He, Y., Gao, W., Jia, Z., Shi, Y., Zhang, S., et al.: Bigdatabench: a big data benchmark suite from internet services. In: HPCA 2014, pp. 488–499. IEEE (2014)

Benchmarking and Workload Characterization

Towards a Big Data Benchmarking and Demonstration Suite for the Online Social Network Era with Realistic Workloads and Live Data

Rui Zhang[1]([✉]), Irene Manotas[2], Min Li[1], and Dean Hildebrand[1]

[1] IBM Research, San Jose, USA
ruiz@us.ibm.com
[2] University of Delaware, Newark, USA

Abstract. The growing popularity of online social networks has taken big data analytics into uncharted territories. Newly developed platforms and analytics in these environments are in dire need for customized frameworks of evaluation and demonstration. This paper presents the first big data benchmark centering on online social network analytics and their underlying distributed platforms. The benchmark comprises of a novel data generator rooted in live online social network feeds, a uniquely comprehensive set of online social network analytics workloads, and evaluation metrics that are both system-aware and analytics-aware. In addition, the benchmark also provides application plug-ins that allow for compelling demonstration of big data solutions. We describe the benchmark design challenges, an early prototype and three use cases.

Keywords: Big data · Benchmarking · Online social networks · Analytics · Spark · Hadoop · Performance · Demo

1 Introduction

The age of online social networks is firmly upon us. Not only have we seen generic online social networks like Facebook and Twitter grow to unprecedented scales, the past few years have also witnessed the miasmic rise of more specialized networks such as LinkedIn (professional network), FourSquare (location) and Instagram (photos). These networks generate terabytes of data per day, host billions of users, and contain a variety of information from message streams to ever-changing graphs. Many online social networks, most notably Twitter, make their data public. It is without surprise that online social networks have become one of the most important sources for data analysis.

The palpable momentum behind big data fuels continuous development in analytics algorithms and large-scale platforms (Apache Hadoop[1], Apache

---

[1] http://hadoop.apache.org.

© Springer International Publishing Switzerland 2016
J. Zhan et al. (Eds.): BPOE 2015, LNCS 9495, pp. 25–36, 2016.
DOI: 10.1007/978-3-319-29006-5_3

Spark[2], Apache Storm[3], etc.). For many analytics, such as graph analysis and user or population modeling, online social networks provide arguably the richest, largest data source available. Given the sheer number of developments in this area, fair and comprehensive evaluation between different hardware, software platforms and analytic algorithms in an online social network context becomes ever more important. End users can leverage such evaluation to navigate (many) competing big data solutions based on their needs; whereas solution providers and developers can measure their solution against the state-of-the-art.

However, there is an unfortunate vacuum when it comes to benchmarking big data platforms and analytics against online social network data. The unprecedented size, variety and ever-changing nature of online social network data put forward unique challenges to data analytics and their underlying data platforms. Most existing big data benchmarks such as BigBench [3] and BigDataBench [11] are designed around data sets and use cases largely different from online social networks. They cannot be used to truly measure new solutions against online social network data. LDBC-SNB [2], a new benchmark very much under development, is the only online-social-network-centric benchmark to our best knowledge. Nonetheless, LDBC-SNB is not specifically geared towards big data and does not consist of workloads readily executable on popular big data platforms such as Spark or Storm. In our humble opinion, both LDBC-SNB and other online-social-network-agnostic benchmarks fall short of meeting the following key requirements of an effective online social network benchmark for big data:

- *Realistic Data.* This requirement is rooted in the Veracity of the famous 5 V model of big data [1]. Most existing benchmarks use simulated data. The simulation is often done based on assumptions that are not necessarily true in real online social networks. For instance, LDBC-SNB simulates friendships based on people's common interests and/or places of study. The simulation would thus miss online social network friendships by being at the same company, by introduction through friends, by accidentally meeting at the same event and so forth.
- *Live Data.* When real data is used in existing big data benchmarks, the data set is often a point-in-time copy and no attempt is made to update the data set. Crucially, online social networks are always changing, as new users join and new content (e.g. tweets, images) are posted. A text sentiment classification solution with fast and accurate results against tweets downloaded last year may not have the same success today, perhaps because there might be more tweets from non-English speaking users that are harder to classify and require more processing. A big data platform that efficiently analyzes the social graph of NYC today, may underperform in a mere few months. As new people move into the city and join the network, its size and structural complexity becomes greater, thus potentially shifting the in-memory vs. on-disk ratio of graph storage as well as the compute demands of the graph.
- *Comprehensive Online-social-network-centric Big Data Workloads.* Online social networks present a fertile ground for a wide range of user-oriented

---

[2] http://spark.apache.org.
[3] http://storm.apache.org.

mining algorithms aiming at inter-connected users and user-generated content. These include text analytics (e.g. topic discovery), image analysis (e.g. principal component analysis), graph analysis (e.g. influencer), trend analysis (e.g. major event predictions), user analytics (e.g. interest or personality analysis) and recommendations (e.g. friend suggestion). These workloads would enable comparison between different platforms (e.g. Spark vs. MapReduce). Existing benchmarks often only cover one or two of these categories. BigDataBench, for instance, only provides graph analytics for online social networks.

– *Analytics-aware Metrics.* Many online social network analytics are machine learning algorithms that build models of classification or prediction. When comparing analytics algorithms, the efficiency of execution is only a part of the story. The accuracy of classification or prediction is equally important. Existing benchmarks only provide metrics at a system or platform level. We are not aware of one that measures classification or prediction accuracy.

It is the goal of our benchmark to become the first benchmark to meet the aforementioned requirements. It can be used to evaluate both platform innovations such as a new caching algorithm for Spark and new analytics algorithms such as a more accurate way to analyze tweet sentiment. The benchmark tracks both live and historic data from Twitter (and other online social networks) and channels it in a way that ensures fair comparison across big data platforms. It aims to provide out-of-the-box implementations of all analytics categories above based on state-of-the-art platforms such as Hadoop MapReduce, Spark and Storm.

The vast amount of new applications made possible by online social networks also open up an unusual opportunity, the opportunity to demonstrate the functionality of a big data platform in a compelling way. Benchmark designs to date have solely focused on testing system efficiency. The result is workloads that differentiate platforms with regard to execution time, throughput etc., not necessarily workloads that can show what exciting applications a big data platform has made possible. We have run into many scenarios where the latter is necessary, in addition to the usual emphasis on efficiency. Hence we have designed the benchmark to provide additional application-level plug-ins that can be used to showcase the functional or business values of big data solutions. For instance, we have provided a stock recommendation application (together with a GUI) that builds on Hadoop-based sentiment analysis.

This paper makes the following contributions:

– We identify key challenges in building a big data benchmark based on online social network data, in terms of data, workloads and metrics.
– We present the design and an early prototype of a benchmark that meets these challenges.
– We put forward the additional notion of demo applications for big data benchmarks.

The following two sections describe the challenges and design of our benchmark. Section 4 presents use cases. Related work is discussed in Sect. 5, before we offer concluding remarks and future plans in Sect. 6.

2 Challenges

To further set the stage for this work, we first describe the major challenges in designing an online social network benchmark for big data, with respect to data generation, workloads and evaluation metrics.

2.1 Realistic and Live Data

Online social networks are constantly evolving as new content is generated and new users are registered. Certain analytics such as graph and text analysis can exhibit vastly different behaviors depending on the text content and graph structure. The benchmark must keep pace with this growing complexity, so as to deliver the most meaningful evaluation of big data systems.

Integrating the latest online social network data into a benchmark is far from simple. Firstly, the benchmark must have a means to obtain the latest data from online social networks. This often means building a crawler that can acquire a sufficient sample from the online social network via carefully identified APIs.

Secondly, one must balance keeping the data up-to-date with making it possible to use the same data set for fair evaluation of different systems. This implies taking reasonable snapshots of the data, as new data is continuously downloaded.

Third, there is an additional challenge in organizing the downloaded data. The organization needs to take into account both time and the snapshot copies above. Different data may need different ways of organization. Text data such as tweets, where data items are relatively independent from one another, may be organized differently from graph data where the data points are (far) more correlated.

Last but not least, steps can be taken to minimize the users efforts in transforming the downloaded data into a storage format they are interested in. For instance, the users may be interested in a specific file format such as a standard row-oriented format or a column-oriented format like Parquet[4], an increasingly popular format supported by many big data platforms. They may also want to place the files in one of several storage systems including HDFS and object stores like OpenStack Swift[5], in which case some file grouping or splitting might be needed to attain optimal (or even acceptable) performance. For instance, HDFS has a known problem in handling a large number of small files, thus some automatic grouping would be useful.

2.2 Comprehensive Workloads

Designing a benchmark like ours imposes a range of challenges regarding the workloads that comprise the benchmark. We focus on the main challenge of offering a comprehensive set of workloads that represent a sufficient range of big data analytics relevant to the online social network domain. This ensures big

---

[4] https://parquet.apache.org/.
[5] https://wiki.openstack.org/wiki/Swift.

data platforms and algorithms can be evaluated with regard to a wide range of realistic use cases.

2.3 Evaluation Metrics

A successful benchmark depends on a well-defined set of metrics for apple-to-apple quantitative comparison between different levels of platform offerings, analytics algorithms, cluster configurations etc. Such differentiation allows users to make an informed decision between solutions based on their needs.

Defining an effective set of metrics for online social networks is challenging. One reason is that there is not a single set of metrics that work for all the workloads we plan to cover. For example, some streaming applications require the underlying data analytic frameworks to support near real time processing capability which might be in the order of milliseconds. On the other hand, for batch workloads such as iterative machine learning, end users may care more about job throughput. Moreover, metrics not directly related to systems performance also come into play. Scalability metrics are needed to show how the performance of the system can change when the data and cluster size scale up and down. In order to evaluate different analytics algorithms (that perform the same type of analysis on the same data), we also need a set of metrics to evaluate the output quality (e.g. classification accuracy) of algorithms being compared.

3 Benchmark Design and Early Prototype

This section presents our benchmark design, as shown in Fig. 1. Herein we describe several key components and their current prototyping states in detail.

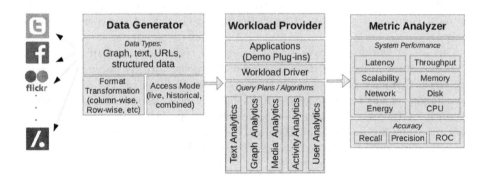

Fig. 1. Architecture

3.1   Data Generator

The benchmark currently uses the Twitter Search API[6] to crawl live tweet-related data. We have chosen Twitter as the primary data source, as it is the largest online social network with an open data policy. The Search API acquires a sample of the latest tweets on Twitter that match given keywords. The crawler is implemented in PHP. It maintains a list of default keywords but can also be customized by the users with regard to their interests.

Each downloaded tweet is a JSON object. We pre-process the data to extract the tweet text, the user profile information, time stamp, any attached image and/or URL. For Twitter, we have not encountered very large objects (e.g. over the 5 GB size supported by OpenStack Swift) that need to be split up. Instead, data associated with tweets are bundled into one file each day by default. One is free to take the latest N days of data for an evaluation. We plan to experiment with different bundle sizes.

We are able to place and extract data from cluster file systems like HDFS and object stores such as OpenStack Swift. The Search API yields around 1 % of the latest tweets. We are exploring means to sample more data and vary the distribution of the samples.

3.2   Workload Provider

We have identified a comprehensive set of workloads that represent different online social network analytics that can be performed on the data collected by the data generator. The comprehensive set of workloads are described in Table 1. In this table, the first column *Source Type*, represents the type of data source used (i.e., plain text files, graphs, etc.); the second column *Category*, shows the category of the data analysis; the third column *Analysis Type*, represents the kind of analysis to be performed on the data; the fourth column *Methods*, lists the most common algorithms or heuristics used to carry out the corresponding analysis.

Our set of workloads cover a broad range of techniques used in the analysis of online social network data. Some workloads are already being integrated into the benchmark using existing Hadoop or Spark libraries. For example, in the category of media analytics the HIPI[7] library provides an image processing interface for Hadoop. For text analytics, different implementations of LDA for topic modeling are available for Hadoop[8] and Spark[9]. The same is true for item recommendations[10,11]. We plan to open up our APIs for third-parties to add new workloads into the benchmark.

---

[6] https://dev.twitter.com/rest/public/search.
[7] http://hipi.cs.virginia.edu/index.html.
[8] https://github.com/lintool/Mr.LDA.
[9] https://github.com/mertterzihan/pymc/tree/pyspark/pymc/examples/lda.
[10] https://mahout.apache.org/users/recommender/intro-als-hadoop.html.
[11] https://mahout.apache.org/users/recommender/intro-cooccurrence-spark.html.

Table 1. Description of workloads

Source type	Category	Analysis type	Methods
Image, audio, video	Media analytics	Metadata scan	SQL-like queries [b,a]
		Feature extraction, Data compresion	Principal Component Analysis (PCA)[b]
Text	Text analytics	Topic modeling	Latent Diricnhlet Allocation (LDA) [b,a]
		Opinion formulation	Sentiment analysis (e.g., with Naive Bayes Classifier)[b,a]
Graph	Graph analytics	Influencer	Centrality[a], Characteristic Path length (CPL), PageRank[b,a]
		Community detection	Graph density[a], Louvain[b]
Graph, text	Activity analytics	Location prediction & Event Prediction	Logistic regressions[b,a] Ensembles (e.g., Random forests, Gradient-boosted trees)[a]
Graph, text, media	User analytics	Collaborative filtering (Recommendations)	Alternating Least Squares (ALS)[b]

[a] indicates implementation is available in Hadoop MapReduce.
[b] indicates implementation is available in Spark.

3.3 Demo Plug-Ins

It is a goal of our benchmark to provide out-of-the-box applications to be used for demonstrating big data solutions. These demo applications will be built on the social network analytics workloads in the benchmark. We believe that it is important not only to demonstrate the efficiency of novel big data systems or algorithms via the benchmark, but also to showcase what exciting applications these systems or features would enable. New system features and analytic algorithms run the risk of being under-appreciated, specially from a business or end-user point of view, if their use cases are not properly highlighted.

In our current prototype, we have included an application that recommends the most promising stock combinations to purchase based on the sentiment being expressed about the corresponding companies on Twitter. The application is based on demonstrated correlation between tweet sentiment and stock market movement [8]. It currently directs the data generator to crawl tweets related to companies in the S & P 100 index. The application currently leverages two workloads: a MapReduce-based sentiment analysis workload which classifies each tweet download in a given period as positive, negative or neutral, and further aggregates the per-tweet classification at a stock and portfolio (stock combination) level; and a query workload that finds the top N portfolios with the most positive sentiment. We are adding a third workload to analyze each tweet live in a streaming fashion (vs. in the batch fashion in the MapReduce workload). The query workload is linked to a GUI, which displays the stock combinations recommended.

We have used the stock recommendation application at several major conferences to demo different big data innovations (see [12] for an example). We have noted that the application helped engage people who would otherwise not pay attention to the features or fail to appreciate their values. The benchmark proposed in this paper is to our knowledge the first benchmark to introduce the notion of functionality demonstration. It is our hope that benchmarking results can be better understood and valued when given a compelling context. We plan to identify and bundle additional applications in the benchmark and provide the facilities (e.g. script wrappers that make benchmark workloads easy to integrate into a GUI) for the community to do so.

3.4   Metrics Evaluator

Within the benchmark, we focus on two types of metrics: system performance metrics and analytic accuracy metrics. While the former concentrates on the latency, throughput, system utilization, power consumption and scalability, the latter focuses on how accurate the data analytics algorithms can conduct prediction or classification.

System performance metrics are common across all the workload types within the benchmark. A tradeoff is often needed between latency and throughput when running online social network applications on data processing frameworks. With a given cluster size and setup, the lower latency and higher throughput the system can offer, the better it is. For streaming applications and interactive queries, we plan to enable users to evaluate and identify the peak throughput the system can achieve by increasing the data set size and varying the data arrival interval based on a user-specified latency constraint. Moreover, system utilization metrics help users pinpoint the performance bottlenecks and compare amongst frameworks. Given a fixed number of machines, a better system would have lower and more balanced resource utilization. In other words, the resource utilization offers deeper insights into why an underlying system can achieve higher throughput while maintaining low latency. Power consumption is an utilization metric that can help vastly reduce the cost of operating a data center. Our benchmark guides users towards big data solutions that consume less power if other metrics are comparable. In addition, the benchmark includes scalability as one of its metrics. A good system should be able to increase the throughput while maintaining low latency as the number of machine increases.

For analytics accuracy metrics, we initially choose confusion matrix, precision, recall, F-measure and Receiver Operator Characteristics (ROC) curves, as defined in [9]. The confusion matrix defines the basic true or false positives and true or false negatives. Precision and recall captures how well the algorithm can detect true positives while minimizing false negatives. On the other hand, ROCs demonstrate, as the number of false negatives varies, how the number of true positives changes accordingly.

4   Use Cases

Our benchmark can be used to evaluate new analytics algorithms, novel big data systems or optimizations done to existing systems. Its evaluation can focus on an entire big data environment or specific sub systems such as data storage. Herein we discuss several use cases, although space constraints preclude us from providing extensive details.

4.1 Evaluating Big Data QoS Impact

The increasing multi-tenant nature of the cloud and big data environments requires differentiation between analytics workloads sharing the same underlying system and competing for resources. In previous work [12], we created a Quality of Service (QoS) capability for big data that maintains relative priority between workloads from Hadoop MapReduce through to back-end storage.

We used the demo application (see Sect. 3.3) in our benchmark to evaluate the QoS feature via a with-vs-without comparison. The QoS feature was first disabled so that one can observe the severity of storage bandwidth contention between (a) a batch analytics workload that scores stock combinations based on their corresponding tweets and (b) an interactive workload that queries the top N stock combinations. Later in the run the QoS feature was enabled to resolve the contention. Our benchmark, in this case, allowed us to observe the % improvement in interactive query response thanks to intelligent control.

This is an evaluation that would not have been possible with any existing big data benchmark reviewed in Sect. 5, because they either lack the means of acquiring real Twitter data or the support for big data workloads required.

4.2 Constructing an Object Store Benchmark

Object storage [5] has been rapidly gaining momentum in recent times, due largely to its unique ability to scale to a large number of small objects and storage-agnostic REST-ful APIs. Running analytics on object storage is a very recent development and there is a clear void in assessing the performance of object storage against these new workloads. Existing object storage benchmarks, such as COSBench[12], are merely simplistic read/write request generators. We are constructing a sub benchmark for object stores using selected benchmark data and analytics workload combinations as shown in Table 2.

Table 2. Sub benchmark for object storage

Data	Workload
Images linked from tweets	PCA (MapReduce)
Web documents linked from tweets	LDA (MapReduce, Spark)

We have chosen linked images and linked web documents as the data set because they often range in KB to MB in sizes and a large number of tweets link

---

[12] https://github.com/intel-cloud/cosbench.

to these data. Currently we rely on the MapReduce implementation of PCA, but hope to include a Spark implementation in the future. Note that no benchmark reviewed in Sect. 5 directly provides image data, which is arguably the data type most suited for object stores as they are small, abundant and unstructured.

4.3   Evaluating Spark Core Optimization

The increasingly wide adoption of Spark has spurred broad interest in optimizing its core system for superior performance. Our benchmark can be used to evaluate the effectiveness of performance optimizations of Spark such as scheduling policies, memory management and caching optimization. In the near future, we plan to use the workloads contained within the benchmark to evaluate the effectiveness of these specific Spark optimizations.

5   Related Work

To the best of our knowledge, this is the first work that proposes a big data benchmark tailored for online social network data and analytics. Table 3 summarizes the most relevant benchmarks [2–4,11]. Although most of these benchmarks (except LDBC-SNB [2]) are presented as capable of evaluating different big data platforms, they lack mechanisms to collect and analyze live data. These

Table 3. Big data benchmarking: state of the art

Benchmark	Target domain	Data generator	Workload types	Metrics
HiBench [4]	Various	Synthetic data, Static real data	Micro-benchmarks, Web search, ML (Bayesian Class., k-means), HDFS (read/write)	System performance (e.g. execution time, throughput, resource utilization)
LDBC-SNB [2]	OLTP, Graph analytics	Synthetic data	Interactive query-based, graph algorithms	System performance (e.g. execution time, acceleration factor, scalability, throughput)
BigBench [3]	Retail systems	Synthetic data	Interactive-based queries	System performance (execution time)
BigDataBench [11]	Internet services (Web search, e-commerce, online social networks)	Synthetic data, Static real data	Micro-benchmarks, query-based, Page-rank, Connected components	System performance (e.g. throughput, MIPS, MPKI)
Our benchmark	Online social networks	Live real data, Static real data	Media, text, graph, activity and user-based analytics	Analytics accuracy (e.g., precision, recall, ROC); System performance (e.g., execution time, throughput)

benchmarks also only support a relatively narrow set of online social network workloads, mostly focusing on graph-analytics or/and micro-benchmarks.

Amongst the reviewed benchmarks, LDBC-SNB and BigDataBench [11] are the most related. LDBC-SNB proposes a social network data generator and an interactive workload for structured and semi-structured data. Unlike our benchmark, LDBC-SNB focuses solely on generation of synthetic data, and offers mostly graph analysis workloads. Note that LDBC-SNB does not seem to provide any benchmarking workloads based on big data platforms such as Spark. BigDataBench is presented as a benchmark that offers workloads for Internet Services such as e-commerce, search engines and social networks. Although Big-DataBench provides several workloads that cover several categories of web services, the number of workloads is small for each category (including the online social network category). Furthermore, BigDataBench adopts a synthetic data approach. In comparison, our benchmark provides a board set of online social network workloads and uses real data.

Literature on data crawling and simulation is also relevant. Studies on capturing and analyzing data from online social networks have shown the evolving nature of this kind of data sources [7], motivating the need to capture data of different sizes, time-frames and in real-time. This is the philosophy that our benchmark tries to follow. By contrast, existing works on data simulation are mainly designed to create synthetic data sets [6, 10] and lack evaluation methods that support the veracity property of the data being generated for benchmarking.

6   Concluding Remarks

Although it is still work-in-progress, our benchmark has taken an important step towards an inaugural online-social-network-centric benchmark for big data. We plan to continue extending the benchmark with regard to its workload coverage and data variety (e.g. Twitter follower graph and data from other public networks). We are conducting extensive experiments to verify the usefulness of the benchmark, and where possible, to quantify the benefits it offers compared to existing benchmarks. It is hoped that more compelling demos will also be added.

References

1. Demchenko, Y., Grosso, P., De Laat, C., Membrey, P.: Addressing big data issues in scientific data infrastructure. In: 2013 International Conference on Collaboration Technologies and Systems (CTS), pp. 48–55. IEEE (2013)
2. Erling, O., Averbuch, A., Larriba-Pey, J., Chafi, H., Gubichev, A., Prat, A., Pham, M.D., Boncz, P.: The ldbc social network benchmark: interactive workload. In: Proceedings of SIGMOD (2015)
3. Ghazal, A., Rabl, T., Hu, M., Raab, F., Poess, M., Crolotte, A., Jacobsen, H.A.: Bigbench: towards an industry standard benchmark for big data analytics. In: Proceedings of the 2013 ACM SIGMOD International Conference on Management of Data, pp. 1197–1208. SIGMOD, ACM (2013)

4. Huang, S., Huang, J., Dai, J., Xie, T., Huang, B.: The hibench benchmark suite: Characterization of the mapreduce-based data analysis. In: 2010 IEEE 26th International Conference on Data Engineering Workshops (ICDEW), pp. 41–51 (2010)
5. Mesnier, M., Ganger, G.R., Riedel, E.: Object-based storage. IEEE Commun. Mag. **41**(8), 84–90 (2003)
6. Ming, Z., Luo, C., Gao, W., Han, R., Yang, Q., Wang, L., Zhan, J.: Bdgs: A scalable big data generator suite in big data benchmarking. In: Rabl, T., Raghunath, N., Poess, M., Bhandarkar, M., Jacobsen, H.-A., Baru, C. (eds.) Advancing Big Data Benchmarks. Lecture Notes in Computer Science, vol. 8585, pp. 138–154. Springer, Heidelberg (2014)
7. Mislove, A., Marcon, M., Gummadi, K.P., Druschel, P., Bhattacharjee, B.: Measurement and analysis of online social networks. In: Proceedings of the 7th ACM SIGCOMM Conference on Internet Measurement, pp. 29–42. IMC, ACM, New York (2007)
8. Oh, C., Sheng, O.: Investigating predictive power of stock micro blog sentiment in forecasting future stock price directional movement. In: Galletta, D.F., Liang, T.P. (eds.) International Conference on Information Systems. Association for Information Systems (2011)
9. Powers, D.M.: Evaluation: from precision, recall and f-measure to roc, informedness, markedness and correlation. J. Mach. Learn. Technol. **2**(1), 37–63 (2011)
10. Rabl, T., Danisch, M., Frank, M., Schindler, S., Jacobsen, H.A.: Just can't get enough - synthesizing big data. In: Proceedings of the ACM SIGMOD Conference (2015)
11. Wang, L., Zhan, J., Luo, C., Zhu, Y., Yang, Q., He, Y., Gao, W., Jia, Z., Shi, Y., Zhang, S., Zheng, C., Lu, G., Zhan, K., Li, X., Qiu, B.: Bigdatabench: a big data benchmark suite from internet services. In: 2014 IEEE 20th International Symposium on High Performance Computer Architecture (HPCA), pp. 488–499 (2014)
12. Zhang, R., Jain, R., Sarkar, P., Rupprecht, L.: Getting your big data priorities straight: a demonstration of priority-based qos using social-network-driven stock recommendation. Proc. VLDB Endow. **7**(13), 1665–1668 (2014)

On Statistical Characteristics of Real-Life Knowledge Graphs

Wenliang Cheng, Chengyu Wang, Bing Xiao, Weining Qian[✉],
and Aoying Zhou

Institute for Data Science and Engineering,
ECNU-PINGAN Innovative Research Center for Big Data,
East China Normal University, Shanghai, China
{wenliangcheng,chengyuwang,bingxiao}@ecnu.cn,
{wnqian,ayzhou}@sei.ecnu.edu.cn

Abstract. The success of open-access knowledge graphs, such as YAGO, and commercial products, such as Google Knowledge Graph, has attracted much attention from both academic and industrial communities in building common-sense and domain-specific knowledge graphs. A natural question arises that how to effectively and efficiently manage a large-scale knowledge graph. Though systems and technologies that use relational storage engines or native graph database management systems are proposed, there exists no widely accepted solution. Therefore, a benchmark for management of knowledge graphs is required.

In this paper, we analyze the requirements of benchmarking knowledge graph management from a specific yet important point-of-view, i.e. characteristics of knowledge graph data. Seventeen statistical features of four knowledge graphs as well as two social networks are studied. We show that through these graphs depict similar structures, their tiny differences may result in totally different storage and indexing strategies, that should not be omitted. Finally, we put forward the requirements to seeding datasets and synthetic data generators for benchmarking knowledge graph management based on the study.

Keywords: Benchmark · Knowledge graph · Social network

1 Introduction

Recent years have witnessed an enthusiasm on the construction and application of large-scale knowledge graphs in both academia and industry communities. A knowledge graph can serve as the backbone of Web-scale applications such as query expansion [6], question answering [11] etc. For example, there are over 500 million entities and 3.5 billion facts in Google Knowledge Graph [18], which is employed to enhance Google search engine's search results. A natural question arises that how to efficiently manage such large-scale knowledge graphs. Though systems and technologies, such as using relational database management systems with specific indexing, or building native graph databases, are developed,

© Springer International Publishing Switzerland 2016
J. Zhan et al. (Eds.): BPOE 2015, LNCS 9495, pp. 37–49, 2016.
DOI: 10.1007/978-3-319-29006-5_4

it is still a research problem. Therefore, a benchmark for knowledge graph management is required for better understanding the problem of knowledge graph construction and utilization, and helping users to select appropriate systems or technologies while using knowledge graphs.

Existing benchmarks for graph data management, such as LinkBenck from Facebook [2], Social Network Benchmark (SNB) from LDBC [8] and our previous work BSMA [16], are mainly designed for social graph management applications. Note that the inherent difference between knowledge graphs and social networks is that the vertices and edges in knowledge graphs are usually typed or annotated with rich attributes or semantic labels, while in social graphs, the number of semantic labels of vertices and edges is limited. We argue that these benchmarks should not be used for benchmarking knowledge graph management, not to mention that they are accessed by different kinds of workload.

In this paper, we analyze the requirements for benchmarking knowledge graph management from a basic yet important point-of-view, i.e. statistical characteristics that depict structures of knowledge graphs. We show that the semantic nature of knowledge graphs results in different characteristics from knowledge graphs to social networks, as well as between different knowledge graphs, and different parts within a knowledge graph.

The contributions of this paper are as follows:

- Thirteen statistics and four statistical distributions are introduced for characterizing structures of graphs. Their intuitive meaning and effectiveness on managing graphs are analyzed.
- Studies over four knowledge graphs, including both common-sense knowledge graphs and domain-specific ones, and two social networks, are conducted. The empirical studies show that though these graphs are similar in certain structure characteristics, such as power-law distribution of vertex degrees, they are different in other features. Detailed analysis on how and why these differences exist is provided.
- The requirements for seeding datasets and synthetic data generator for benchmarking knowledge graphs are analyzed. We show that the synthetic data generator cannot be trivially adapted to an existing social network generator.

The rest of this paper is organized as follows. In Sect. 2 we introduce the related works on statistical characteristics and benchmarks in large scale graph. In Sect. 3, we introduce the statistical metrics used to evaluate the graphs. In Sect. 4 we describe four knowledge graphs and two social networks we experiment on. In Sect. 5, we present our empirical studies on this issue. Finally, we conclude with a summary and propose the future work in Sect. 6.

2   Related Work

A benchmark for knowledge graph management requires a clear understanding of the statistical characteristics of knowledge graphs. Research works which are focused on analyzing graph structural properties such as complex network, have proposed structural metrics and distributions such as *node degree, hop* and *diameter* for modelling the structural properties of a graph. Benchmarks on big graph systems benefit a lot from them. For example, a benchmark generates synthetic data according to the distributions or topology of a graph. In this section, we give a brief introduction to these research fields.

For the past years, researchers have devoted themselves to analyzing structural properties of large scale graphs. Broder et al. [4] study the web as a graph via a series of graph structural metrics such as *diameter, nodes* and *degree*. For social network, Kumar et al. [12] study the evolutions of Flickr and Yahoo! 360 by analyzing their dynamic time-graph's structure properties, for example, *diameter, degree, community size*, etc. Boccaletti et al. [3] survey the studies of the structure and dynamics of complex network.

At the same time, benchmarks for big graph analytical have also been developed rapidly in recent years. Lancichinetti et al. [13] propose a benchmark for graphs, which pays attention to the heterogeneity in the distributions of node degrees and community sizes. As for social networks, Linkbench [2] characterizes the Facebook graph workload and constructs a realistic synthetic benchmark. Social Network Benchmark (SNB) from LDBC [8] models a synthetic social network which is similar to Facebook. It is the first LDBC benchmark based on the *choke-point* analysis [8], which identifies the technique challenges to evaluate in a workload. BSMA [16] is another applicable benchmark aimed to analyze the social media data based on Sina Weibo (microblog) in China.

The research works of the graph structural properties and benchmarks inspire us to have a close observation of the knowledge graphs' characteristics in order to study the problems of designing a benchmark for knowledge graphs from the point of statistical characteristics.

3   Statistical Characteristics

Both knowledge graphs and social networks can be modelled as a directed graph $G = (V, E)$, where V is the set of nodes (entities or users) and E is the set of directed edges (semantic relations). The thirteen statistics are illustrated in Table 1 and the four distributions are introduced as follows:

Degree Distribution. The degree distribution of G is $p(d) = \frac{n_d}{|V|}$, where n_d is the number of nodes whose degree is d and $|V|$ donates the number of nodes in G. In many graphs, the degree exhibits a *power-law* distribution [5] which has the form: $p(d) \propto L(d)d^{-\alpha}$, where $\alpha > 1$ and $L(d)$ is a slowly varying function. We study in-degree and out-degree separately in this paper.

Distribution of Hops. For a path $P = \{v_1, v_2, ..., v_h\}$ in G. The hop of a path P is defined as $Hops(P) = h - 2$, where h is the number of nodes in P. The distribution of hops reflects the connectivity cost inside a graph.

Distribution of Connected Components. There are *strongly* and *weakly* connected component in graph theory. A strongly connected component is a

Table 1. Description of statistical characteristics

Statistics	Description
#Nodes	Number of nodes
#Edges	Number of edges
#Density	The sparsity of a graph, which is formulated as $D(G) = \frac{|E|}{|V|(|V|-1)}$
#ZIDNodes	Number of nodes with zero in-degree
#ZODNodes	Number of nodes with zero out-degree
#BiDirEdges	Number of bidirectional edges
#CTriads	Number of closed triangles. A closed triangle is a trio of vertices each of which is connected to both the other two vertices.
#OTriads	Number of open triangles. An open triangle is a trio of vertices each of which is connected to at least one of the other two vertices.
AvgCC	Average clustering coefficient. The average clustering coefficient of a graph is defined as $C = \frac{3 \times \#Closed\ triads}{\#Open\ triads}$ [19].
FMWcc	Fraction of nodes in max weakly connected component
FMScc	Fraction of nodes in max strongly connected component
AppFdiam	Approximately full diameter
90%EffDiam	The 90 percentile effective diameter, measures minimum number of hops in which 90% of all connected pairs of nodes in a graph are reachable.

community in which any pair of nodes are reachable. A weakly connected component is a set of nodes in which any two nodes are reachable regardless of the edges' direction. The connected components reflects the connectivity of a graph.

Distribution of Clustering Coefficient. The definition of the clustering coefficient w.r.t. a node v_i is $C_i = \frac{|\{e_{jk} : v_j, v_k \in N_i, e_{jk} \in E\}|}{|N_i|(|N_i|-1)}$ [20], where e_{jk} is the edge between v_j and v_k $(j \neq k)$ and N_i is the set of neighbour nodes of v_i. The clustering coefficient measures the nodes' tendency to cluster together.

4   Data Description

In this section, we describe two social networks and four knowledge graphs we study in this paper.

Social Network. Sina Weibo is a famous social media which provides micro-blog service in China. In this paper, we generate two graphs consisting of *persons* and *fellowship* relations. In the first graph, we generate 0.2 million users **randomly** (SNRand) from the entire user set, which have more than 5 million relations between them. In the second graph, we select 0.2 million **most active users** including 36 million edges among them (SNRank). Note that the graphs are neither synchronized nor complete. However, as the comparison to knowledge graphs, the SNRand can simulate the real-life data and SNRank can simulate the most critical situation where activities in social networks are very intense.

WordNet [9] is a lexical network for the English language designed in Princeton University. In WordNet, English words are grouped into sets of cognitive synonym (e.g. nouns, verbs, adjectives and adverbs), in which every synonym stands for a distinct concept. We utilize the real-life *WordNet* directly in our experiment. The words or concepts are nodes and semantic relations are edges.

YAGO2 [10] is a huge semantic knowledge graph which harvests knowledge from *WordNet*, *Wikipedia* and *GeoNames*[1]. We generate three subgraphs from YAGO2, named *YagoTax*, *YagoFact* and *YagoWiki*. *YagoTax* is the taxonomy of YAGO2, consisting of *subClassOf* relations and reflecting the taxonomic knowledge. *YagoFact* contains all the *factual* relations of YAGO2, standing for the factual knowledge. *YagoWiki* consists of the hyperlink relations (*linkedTo*) in YAGO2, reflecting the natural hyperlink structure of Wikipedia.

DBpedia [14] is a multi-language knowledge base extracted from the Wikipedia. The English version of DBpedia describes 4.58 million entities and 2,795 different properties. It utilizes the mapping-based technique to extract facts from Wikipedia info-boxes. We generate the synthetic data *DBpediaFact* from all the factual knowledge of DBpedia.

Enterprise Knowledge Graph(EKG) is a *domain specific knowledge graph* constructed by us. It models the relationships among people, companies, products for customer relation management (CRM). We extract relations from 2 million news articles from Sina Finance News[2] using a bootstrapping strategy similar to *snowball* [1] to iteratively detect relation tuples from entities.

Note the fact that all the knowledge graphs data we generated are real-life, which makes our empirical studies on these graphs more convincing.

5   Empirical Studies

In this section, we evaluate the graphs with a toolkit SNAP [15] and conduct an association rule mining experiment to conduct a series of empirical studies.

5.1   Analysis for Statistics

We analyze the graphs in three groups based on different objectives: 1) In order to study the different aspects of a knowledge graph, we compare the three subgraphs of YAGO2 and make an in-depth analysis of them. 2) We take the four knowledge graphs into consideration and make a series of horizontal comparisons, trying to reveal the differences between each knowledge graph in detail. 3) We compare the knowledge graphs with social networks, attempting to explain why and how the

---

[1] http://www.geonames.org/.
[2] http://finance.sina.com.cn/.

differences exist between them. The evaluation results are listed in Table 2. Those statistics are normalized by #Nodes in order to make them be comparable. Notice that the symbol # before the statistics in Table 1 are replaced by % in Table 2 and only #Nodes and #Edges are retained.

Table 2. Normalized statistics of graphs

Statistics	YagoTax	YagoFact	YagoWiki	DBpedia	WordNet	EKG	SNRand	SNRank
#Nodes	4.49e+5	2.14e+6	2.85e+6	4.26e+6	9.79e+4	9.45e+3	2.00e+5	2.02e+5
#Edges	4.51e+5	3.99e+6	3.80e+7	1.44e+7	1.54e+5	1.21e+4	5.45e+6	3.68e+7
Density	2.02e-6	1.75e-6	9.38e-6	1.59e-6	3.21e-5	2.72e-4	2.72e-4	1.80e-3
%ZIDNs	0.958	0.706	0.184	0.461	0.056	0.240	0.128	0.003
%ZODNs	5.78e-5	0.215	0.010	0.198	0.492	0.515	0.010	0.011
%BDEdges	0.000	0.019	2.940	0.129	0.487	0.498	6.984	81.29
%CTriads	0.000	0.365	26.02	2.115	0.043	0.093	59.92	2,167
%OTriads	2,982	93.62	616.9	371.4	30.66	14.82	5.94e+4	2.26e+5
AvgCC	0.000	0.095	0.331	0.325	0.032	0.029	0.105	0.067
FMWcc	0.998	0.953	0.999	0.989	0.988	0.655	1.000	1.000
FMScc	0.000	0.006	0.778	0.051	0.204	0.162	0.854	0.985
AppFdiam	11.00	15.00	14.00	40.00	25.00	18.00	15.00	7.000
90 %EDiam	6.740	5.340	3.830	5.920	10.800	6.770	5.090	3.350

Comparison Between YAGO2's Subgraphs. The three subgraphs *Yago-Tax*, *YagoFacts* and *YagoWiki* describe three aspects of YAGO2 respectively. From Table 2 we can see, the %CTriads and %OTriads of the three subgraphs are different significantly. The %OTriads of *YagoTax* is highest while the %CTriads is 0.00. In *YagoFacts* and *YagoWiki*, the differences between %CTriads and %OTriads are relatively smaller. The AvgCC of *YagoWiki* is higher than *YagoTax* and *YagoFacts*, indicates that the nodes in *YagoWiki* are more likely to be clustered via the relation *linkedTo* than the taxonomic relation *subClassOf* in *YagoTax* and *factual* relations in *YagoFact*. The relative differences of the three subgraphs' %ZODNs are greater than that of %ZIDNs in general, shows that the *out-degree* distributions of the three subgraphs are more diverse than their *in-degree* distributions.

Comparison Between Knowledge Graphs. To make the comparisons among the four knowledge graphs meticulously, we further divide them into two groups according to their contents: **taxonomic level** (e.g. *YagoTax* and *WordNet*) and **factual level** (e.g. *YagoFact*, *DBpediaFact* and *EKG*).

In **taxonomic level** group, the %ZIDNs and %ZODNs of *YagoTax* show almost every node in *YagoTax* has *out-degrees* while 95.8 % of the nodes have no *in-degree*. Because the *YagoTax* is a taxonomy tree with few hierarchies and tremendous unconnected leaves. In *WordNet*, the FMScc shows that 20.4 %

of the nodes are in the max *strongly connected component* and the *%ZODNs* shows half of the nodes in *WordNet* have no *out-degree*. We imply the topology of *WordNet* is a star structure with a max *strongly connected component* (20.4 % nodes) in the centre and the other half of nodes (49.2 %) are distributed outside.

In **factual level** group, the *YagoFact* and *DBpediaFact* are both *common-sense knowledge graphs* (CSKG) and *EKG* is a *domain-specific knowledge graph* (DSKG). We compare *YagoFact* and *DBpediaFact* first. The scale and *density* of *DBpediaFact* are greater than *YagoFact*, indicates that the automatically generated *DBpediaFact* [14] contains more entities than them in *YagoFact*, where the relation extraction method is based on hand-written rules [10]. The *%CTriads* and *%OTriads* show nodes in *DBpediaFact* are more likely to form *triangles* than *YagoFact*. As for the *CSKG* (*YagoFact*, *DBpediaFact*) and *DSKG* (*EKG*), the *density* of *DSKG* is higher than *CSKG*. The *%CTriads*, *%OTriads* as well as *AvgCC* show that the *DSKG* have more *triangles* than *CSKG*, too.

Comparison Between Knowledge Graphs and Social Networks. The *density* shows that social networks are denser than knowledge graphs. Due to the activeness of people, social networks contain more *bidirected* edges than knowledge graphs. The *%CTriads* and *%OTriads* of social networks are greater than knowledge graphs. Because in social networks, a person's friends tend to be friends due to the *triadic closure* property [7]. The *FMWcc* and *FMScc* of social networks show that most of nodes in social networks are in a max *strongly connected component*, while in knowledge graphs, most of the nodes tend to form *strongly connected components* within a small range. Another evidence from *FMWcc* shows that the nodes in knowledge graphs are connected by a max *weakly connected component*. We conclude that the *strongly connected components* in knowledge graphs are connected by a series of *bridges* [19] (also known as cut-edge) actually. In short, there exist gaps between the *strong connected components* in knowledge graphs, while the social network is a whole *strong connected component*.

Conclusion. Table 3 summarizes the statistics which have significant different performances in the three comparison groups. As we can see, the differences between the subgraphs of YAGO2 in the first group are mainly embodied in (*open or close*) *triangle*, *clustering co-efficient* and *strongly connected component* three aspects. Then we can not treat the knowledge graph as a whole graph when generating the synthetic data. We should generate it separately according to different parts or semantic topics.

In the second group, the *density*, *%CTriads*, *%OTriads*, *AvgCC*, *%ZIDNs* and *%ZODNs* are the prominent statistical characteristics which perform diversely between each knowledge graphs on both *taxonomic* and *factual* levels. It implies that when generating synthetic data or designing workloads for a knowledge graph in a special domain, the data characteristics should be considered first and the workloads should emphasize these characteristics in special.

Table 3. The prominent statistics which are different in each comparison group

Experiment groups		Density	%CTriads	%OTriads	AvgCC	FMScc	%BDEdges	%ZDNs
YAGO subgraphs			√	√	√	√		
KGs	Taxonomic	√	√	√				√
	Factual	√	√	√	√			
KGs and SNs		√	√	√		√	√	

[1] %ZDNs donates for %ZIDNs and %ZODNs.
[2] KGs and SNs are short for the "Knowledge Graphs" and "Social Networks".

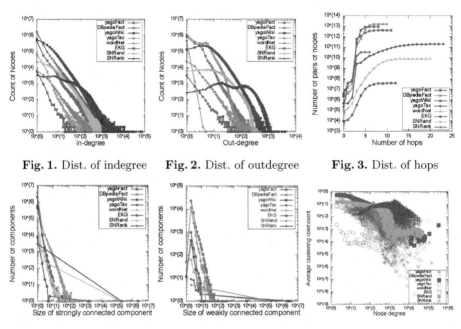

Fig. 1. Dist. of indegree   **Fig. 2.** Dist. of outdegree   **Fig. 3.** Dist. of hops

Fig. 4. Dist. of SCC   **Fig. 5.** Dist. of WCC   **Fig. 6.** Dist. of ACC

In the last group, the *density, triangle, strongly connected component* and *%BDEdges* are the main characteristics which perform different between knowledge graphs and social networks. There are few bidirectional edges in knowledge graphs (but not none). Thus data generator should control their existences appropriately in knowledge graphs. The gaps between *strong connected components* in knowledge graphs remind us that more facts should be extracted and added to the existing knowledge graphs to bridge them in the future. And with the development of them, the benchmarks for knowledge graph data management techniques should focus on different properties dynamically.

5.2 Analysis for Distributions

In this section, we give an in-depth analysis of the graphs' distribution metrics. The distributions are illustrated in Figs. 1, 2, 3, 4, 5 and 6.

Table 4. Fitted parameters for all the distributions

Graphs	InDeg		OutDeg		Hop			SCC		WCC	
	$L(d)$	α	$L(d)$	α	a	b	c	$L(d)$	α	$L(d)$	α
yagoFact	2.07e+5	1.859	4.80e+5	2.245	4.00e+12	5.71	1.22	2.66e+5	4.940	2.13e+5	3.228
YagoTax	1.70e+4	1.855	4.48e+5	8.412	1.95e+11	5.45	0.47	-	-	1.98e+1	1.063
yagoWiki	3.92e+6	1.914	5.61e+8	3.000	9.40e+12	7.84	1.89	1.20e+5	3.849	1.48e+2	0.597
DBpedia	6.77e+5	1.827	2.08e+14	7.697	1.58e+13	10.62	1.96	1.30e+5	2.361	1.55e+5	3.769
WordNet	2.86e+4	2.455	7.86e+4	2.379	7.60e+10	21.40	2.03	7.63e+3	2.745	2.54e+2	2.045
EKG	1.14e+3	1.913	2.96e+2	2.719	3.64e+7	7.49	1.31	7.41e+3	5.236	1.39e+4	3.834
SNRand	5.27e+5	1.884	2.44e+5	1.441	3.55e+10	10.29	3.20	2.85e+4	6.555	-	-
SNRank	1.95e+5	1.293	3.12e+7	2.078	3.43e+10	14.39	4.72	-	-	-	-

[1] The "-" represents that the distribution's points are not enough to fit the parameter.
[2] Due to the divergency of the points, the *average clustering co-efficient* distributions are not fitted.

In Fig. 1, the *in-degree* distributions of all knowledge graphs and social networks exhibit the power-law distributions. The estimated parameters are detailed in Table 4. Nearly all the exponents α are consistently around (1.8,2.4) except that of *SNRank*, which is different with all the others. The initial segment of *SNRank* distribution deviates from the power law greatly until the *in-degree* increases up to around 560.

The *out-degree* distributions are shown in Fig. 2. As we can see, there exist not only significant distinctions between knowledge graph and social network but also between knowledge graphs. The *out-degree* distributions of the three YAGO2 subgraphs are different significantly, which is consistent with the analysis in previous section. All the distributions are deviated from power law initially, and they are diverse with each other as well. The descent rates of the distributions also vary widely. The fitted parameters α in Table 4 fluctuate from 1.4 to 8.4.

Figure 3 illustrates distributions of *hops* in those graphs. All the distributions are in "S" shape. In order to fit the data to some curve, we introduce a variant of *sigmoid* function with the form $f(x) = \frac{a}{1+e^{b-cx}}$. The parameters are fitted very well in Table 4. The max hops of *yagoTax* and *WordNet* are larger than the others in general. The *SNRand* and *SNRank* have the minimal max *hops*, close to 6, which is in consistent with *six degrees of separation* theory [20] in social networks. Another interesting discovery is that with the hop added from 2 to 3, all the distributions increase explosively in general.

Figures 4 and 5 reflect the distributions of *connected components*. Both the distributions of *strongly* and *weakly connected components* of knowledge graphs are in power-law distribution uniformly except *yagoTax* which is a flat tree with a lot of unconnected leaves. While in *SNRand* and *SNRank* as social networks, there only have one max *strongly connected component* and a small part of isolated nodes. The pow-law distributions of *strongly* and *weakly connected components* in knowledge graphs show that the nodes in knowledge graphs are clustered in several small ranges (actually most of the *strongly connected components* are connected by *bridges* according to our analysis in previous section). However, nodes in social networks are organized into one big *strongly connected component*.

Figure 6 presents the distributions of *average clustering coefficient (ACC)*. In this experiment setting, we treat the graphs as undirected and the *degree* is the total of *in-degree* and *out-degree*. Figure 6 shows, all the other graphs are trend to perform the power law initially except *SNRank*, with the value of x increases, the curves start to diverge. The *ACC* of social networks is higher than knowledge graphs in general, which also reflects that the *local clustering property* of knowledge graphs is not as strong as social networks.

Table 5. Relatedness of *in* and *out* relations in YAGO2

InRelation1	InRelation2	$R(r1, r2)$	OutRelation1	OutRelation2	$R(r1, r2)$
rdf:type	subClassOf	0.9948	rdf:type	hasWikipediaUrl	0.9999
playsFor	isAffiliatedTo	0.9797	linksTo	hasWikipediaUrl	0.9894
hasChild	isMarriedTo	0.9182	rdf:type	linksTo	0.9815
wasBornIn	isLocatedIn	0.8712	exports	imports	0.9220
graduatedFrom	worksAt	0.7545	playsFor	isAffiliatedTo	0.7668
created	directed	0.7456	imports	dealsWith	0.7475
actedIn	created	0.6943	exports	dealsWith	0.7454
diedIn	wasBornIn	0.6730	imports	hasTLD	0.6553
wroteMusicFor	directed	0.6041	hasTLD	dealsWith	0.6334
isCitizenOf	dealsWith	0.5590	isConnectedTo	hasAirportCode	0.5984

Conclusion. Figures 1 and 2 illustrate that the *in-degree* and *out-degree* distributions of knowledge graphs and social networks are of great differences. The initial segment of the *out-degree* distributions follow a different kind of distributions (e.g. *poisson* or a combination of *poisson* and *power-law*). Users can generate the synthetic data sectionally according to the different *in* or *out degrees*. Figure 3 shows the distributions of *hops* that exhibit the "**S**" shape distribution, which fits a *sigmoid* function very well. Users can utilize the *sigmoid* function to generate the data. Figures 4 and 5 illustrate that the knowledge graphs are separated naturally into a number of *strongly connected components* and isolated *weakly connected components*, in which the size of the components displays power law distributions. The natural partition ability of the knowledge graph allows us to manage the data distributively, which will potentially reduce the cost of *join* operations significantly. However, in social networks, users should divide them by some special graph partition algorithms. Figure 6 shows the distributions of *average clustering co-efficient* are in the power law distribution, but with the incasement of *degree*, the *clustering co-efficient* starts to diverge, indicating the data generator should not only obey the power law distribution in overall but also embody the data divergence at the same time.

5.3   Analysis for Labels' Relatedness

We conduct an association rule experiment on semantic labels of YAGO2 as a case study to discover the relatedness between labels.

For a relationship, there are two kinds of relations, namely, *out* and *in* respectively. To compute the relatedness between labels r_i and r_j, we first define the support of $r_i \rightarrow r_j$ as $supp(r_i \rightarrow r_j) = \frac{|r_i \cap r_j|}{|r_i|}$, which is inspired by the definition in [17], where $|r_i|$ denotes the number of nodes that have a relation r_i and $|r_i \cap r_j|$ denotes the number of nodes that have both relations r_i and r_j. Obviously, the *support* function is not symmetric, inspired by the definition of F-measure, the relatedness of r_i and r_j is defined as $R(r_i, r_j) = \frac{2 \times supp(r_i \rightarrow r_j) \times supp(r_j \rightarrow r_i)}{supp(r_i \rightarrow r_j) + supp(r_j \rightarrow r_i)}$.

Table 5 lists top-10 pairs of relations with highest relatedness. From Table 5, we find the semantic relations are topic related. For example, *hasChild* and *isMarriedTo* indicate children and marriage belong to the same topic. Some semantic relations have no intersection with others. For example, the relatedness between *hasGender* and *isLocatedIn* is 0.00. However, some relations (e.g. *rdf:type*) almost have co-occurrence with any other relations, which indicates the semantic labels are distributed differently due to the semantics. Thus, in the point of designing benchmarks, we conclude the data generator for knowledge graphs cannot be trivially adapted to an existing social network generator. Note that in many information extraction systems, new kinds of relations should be extracted easily by many automatic methods, for example, the *snowball* [1] system learns new generated patterns to discover new tuples.

6   Conclusion and Discussion

In this paper, we observe four knowledge graphs and two kinds of social networks closely on their statistical characteristics. After our in-depth analysis on the experiments, it is shown that:

(1) Different parts of a knowledge graph have different properties in some certain statistical characteristics, such as *clustering co-efficient, strongly connected component.*

(2) The different types of knowledge graphs have different properties in several statistical characteristics, and their data distributions are different either, such as *out degree distributions* and *hops distributions.*

(3) Knowledge graphs are different with social networks in several distributions, such as *strongly* and *weakly connected components.* With their development, new kinds of relationships could be discovered easily.

These empirical observations show that the existing synthetic graph generators can not generate the real-life knowledge graph data. Therefore, to build a benchmark for testing and evaluating the different knowledge graph management systems and techniques, an important task is to build generators that meet the following requirements:

- The generator should generate synthetic data of a knowledge graph in different aspects, such as *taxonomic knowledge* and *factual knowledge*.
- The generator should take the semantic labels in knowledge graphs into consideration and preserve the statistical characteristics of the real-life data.
- The generator should be able to not only generate the static synthetic data of a certain knowledge graph, but also the different stages of knowledge graph's construction.

The empirical study conducted in this paper is the first effort on modelling statistical characteristics of knowledge graphs. Our future work includes the design and implementation of the data generators of knowledge graphs.

Acknowledgement. This work is partially supported by National Hightech R&D Program (863 Program) under grant number 2015AA015307, and National Science Foundation of China under grant numbers 61432006 and 61170086. The authors would also like to thank Ping An Technology (Shenzhen) Co., Ltd. for the support of this research.

References

1. Agichtein, E., Gravano, L.: Snowball: extracting relations from large plain-text collections. In: DL 2000, pp. 85–94. ACM (2000)
2. Armstrong, T.G., Ponnekanti, V., Borthakur, D., Callaghan, M.: Linkbench: a database benchmark based on the facebook social graph. In: SIGMOD, pp. 1185–1196. ACM (2013)
3. Boccaletti, S., Latora, V., Moreno, Y., Chavez, M., Hwang, D.-U.: Complex networks: structure and dynamics. Phys. Rep. **424**(4–5), 175–308 (2006)
4. Broder, A., Kumar, R., Maghoul, F., Raghavan, P., Rajagopalan, S., Stata, R., Tomkins, A., Wiener, J.: Graph structure in the web. Comput. Netw. **33**(1–6), 309–320 (2000)
5. Clauset, A., Shalizi, C.R., Newman, M.E.J.: Power-law distributions in empirical data. SIAM Rev. **51**(4), 661–703 (2009)
6. Dalton, J., Dietz, L., Allan, J.: Entity query feature expansion using knowledge base links. In: SIGIR, pp. 365–374. ACM (2014)
7. David, E., Jon, K.: Networks, Crowds, and Markets: Reasoning About a Highly Connected World. Cambridge University Press, New York (2010)
8. Erling, O., Averbuch, A., Larriba-Pey, J., Chafi, H., Gubichev, A., Prat, A., Pham, M.-D., Boncz, P.: The ldbc social network benchmark: interactive workload. In: SIGMOD, pp. 619–630. ACM (2015)
9. Fellbaum, C. (ed.): WordNet: an electronic lexical database. MIT Press, Cambridge (1998)
10. Hoffart, J., Suchanek, F.M., Berberich, K., Weikum, G.: Yago2: a spatially and temporally enhanced knowledge base from wikipedia. Artif. Intell. **194**, 28–61 (2013)
11. Joshi, M., Sawant, U., Chakrabarti, S.: Knowledge graph and corpus driven segmentation and answer inference for telegraphic entity-seeking queries. In: EMNLP, pp. 1104–1114. ACL (2014)
12. Kumar, R., Novak, J., Tomkins, A.: Structure and evolution of online social networks. In: KDD, pp. 611–617. ACM (2006)

13. Lancichinetti, A., Fortunato, S., Radicchi, F.: Benchmark graphs for testing community detection algorithms. Phys. Rev. E **78**(4), 046110 (2008)
14. Lehmann, J., Isele, R., Jakob, M., Jentzsch, A., Kontokostas, D., Mendes, P.N., Hellmann, S., Morsey, M., van Kleef, P., Auer, S., Bizer, C.: DBpedia: a large-scale, multilingual knowledge base extracted from wikipedia. Semant. Web J. **6**(2), 167–195 (2015)
15. Leskovec, J., Sosič, R.: SNAP: a general purpose network analysis and graph mining library in C++, June 2014. http://snap.stanford.edu/snap
16. Ma, H., Wei, J., Qian, W., Yu, C., Xia, F., Zhou, A.: On benchmarking online social media analytical queries. In: GRADES, p. 10 (2013)
17. Rajaraman, A., Ullman, J.D.: Mining of Massive Datasets. Cambridge University Press, New York (2011)
18. Singhal, A.: Introducing the knowledge graph: things, not strings. Official Google Blog, May 2012
19. Wasserman, S., Faust, K.: Social Network Analysis: Methods and Applications, vol. 8. Cambridge University Press, New-York (1994)
20. Watts, D., Strogatz, S.: Collective dynamics of 'small-world' networks. Nature **393**, 440–442 (1998)

Mbench: Benchmarking a Multicore Operating System Using Mixed Workloads

Gang Lu[1]([✉]), Xinlong Lin[2], and Runlin Zhou[3]

[1] Institute of Computing Technolgy, Chinese Academy of Sciences,
University of Chinese Academy of Sciences, Beijing, China
`lugang@ict.ac.cn`
[2] Beijing Academy of Frontier Science and Technology, Beijing, China
`linxinlong@mail.bafst.com`
[3] National Computer Network Emergency Response Technical Team,
Coordination Center of China, Beijing, China
`zhourunlin@cert.org.cn`

Abstract. Existing multicore operating system (OS) benchmarks provide workloads that estimate individual OS components or run in solo for performance and scalability evaluation. But in practice a multicore machine usually hosts a mix of concurrent programs for better resource utilization. In this paper, we show that this lack of mixed workloads in evaluation is inadequate at predicting real-world behavior especially in the spectrum of big data and latency-critical workloads.

We present *Mbench*, a novel benchmark suite that helps reveal information about the performance isolation provided by an OS. It includes not only micro benchmarks of least redundancy, but also real applications with dynamic workloads and large data sets. All the benchmarks are integrated into an experiment control tool with components of experiment setup, controlling, monitoring,and analyzing. And further extensions can be applied for supports of more benchmarks. Using the benchmark suite, we demonstrate the importance of considering mixed workloads in some challenging problems ranging from workload consolidation and tail latency performance. We plan to release Mbench under a free software license.

1 Introduction

The purpose of an operating system (OS) is to manage computer hardware and software resources and provide common services for applications. With the increasing resource density in a machine, an OS becomes more capable of provisioning resources for multiple applications. The deployment of applications in industry has thus changed a lot, while the methods we are using in evluating an OS do not. In this paper, we argue that the gap between what an OS should do in practice and which properties we currently measure about becomes large. We show how traditional benchmarks fail to reveal some hidden problems in the evaluation of running a single workload, and propose a benchmark suite so as to evaluate how well an OS can accommodate multiple applications.

© Springer International Publishing Switzerland 2016
J. Zhan et al. (Eds.): BPOE 2015, LNCS 9495, pp. 50–63, 2016.
DOI: 10.1007/978-3-319-29006-5_5

Hardware vendors are continusly making efforts on machines with more processors and cores. Intel Coporation has announced their 18-core processor (Xeon Haswell EP) which attracts many companies in constructing new machines with up to 64 cores. This will much increase the potential to host a large number of programs within a server. In fact, real workloads usually consist of multiple concurrent programs in large scale clusters [7]. However, the benchmarking methodologies still stay on evaluating performance, scalability, or security for a single application. In other words, current multicore OS benchmarks do not evaluate performance isolation between independent apps. For example, lmbench [26] and hbench [13] are micro benchmarks that test the performance of individual components. Newer benchamark suites like BigDataBench [33] and CloudSuite [19] include application benchmarks in terms of distributed computing and are usually used to evalute the architectural functionalities by running them in solo mode. Actually, [21,24,29] have mentioned this problem. But no significant progress is yet made.

As the evaluations are alway conducted in a solo mode, the interactions between competing multithreaded workloads are poorly understood. Moreover, the observations gained by evaluation of running an application in solo may become inaccurate because both performance isolation and scalability are affected by mixed workload. In Sect. 2.3, we show that this lack of mixed workloads in evaluation is inadequate at predicting real-world behaviors especially in the spectrum of big data and latency-critical workloads. Since the OS designs are continuing its steps [9,16,25], it is time we face the challenges.

In this paper, we present a new benchmark suite designed for evaluating the working of an OS using mixed workloads. It is made to acquire the properties necessary for an ideal benchmark suite: (1) covering both micro and application benchmarks; (2) keeping as least redundancy as possible; (3) including latency critical workloads with the metric of tail latency [18] as a special case that needs to be concerned. To reflect real-world scenarios with real applications and dynamic workloads, we include partial benchmarks from BigDataBench. Through this benchmark suite, we provide the guidance to address the challenges of choosing an appropriate mix for consolidation by surveying the candidate workloads in resource utilization, and an extreme mix for exposing the potential bottlenecks. Besides, we develop an experiment controlling tool (*Mcontroller*) to faciliate the supervising of experiments of mixed workloads. All the benchmarks are integrated into the tool with components of experiment setup, controlling, monitoring, and analyzing. Besides, further extensions can be applied for supports of more benchmarks.

In the next section we review the benchmarking methodologies used in OS research papers in the past 20 years, and that such benchmarks fail to capture important aspects of how an OS manages resources under a mixed workload in the terms of big data and latency critical applications. We then introduce the new benchmark suite and the experiment control tool for mixed applications. After that, we present some use cases in evaluating the performance isolation of operating systems.

2   Current Approaches

In this section, we survey popular OS benchmark suites in the research community and benchmarks used by OS researchers in the recent twenty years. And we show how existing benchmarking methodologies fail to expose issues in mixed scenarios.

2.1   Popular OS Benchmark Suites

The Unix-based operating systems have evoled over thirty years, but there are few works about OS benchmarking. Most of the benchmark suites focuses on benchmarking subsystems of an OS using microbenchmarks, including programs written with special contexts and kernel algorithms. With the emergence of multicore machines, OS benchmarks begin to include evaluations of the scalability of an OS with the increasing number of cores.

Table 1 lists some benchmark suites widely used by OS researchers, especially lmbench and hbench. lmbench and hbench are composed of portable benchmarks designed to measure the performance of primitive functionality provided by an operating system or hardware platform. They are ideal for both performance measurement and analysis in a general purpose OS. BenchIT is designed for high performance computing (HPC) and composed of microbenchmarks and four MPI applications. These suites played great roles in evaluating performance of

Table 1. Some popularly used OS benchmark suites. (Benchmarks in *italic* are application benchmarks).

Year	Benchmark suites	Properties	Benchmarks
1983	unixbench [3]	performance	string handling; scientific applications; exec; file copy; pipe throughput; process creation; shell; system call; graphical tests
1996	lmbench [26]	performance	Bandwidths: cache; pipe; tcp; Latencies: context switch; network; file system; process creation; signal handling, syscall overheads; memory latency; Miscellanious: Processor clock rate calculation
1997	hbench [13]	performance	cache behavior; memroy bandwidth; process creation; file access
2004	BenchIT [22]	performance for HPC	memory (bandwidth &latencies); file access; MPI communication; database (MySQL); sting operations; sort algorithms; binary search; numerical algorithms; *applications (CGV; iRODS; MGV; reflection)*
2010	mosbench [12]	scalability	*Exim; memcached; Apache; PostgreSQL; gmake; Psearchy; MapReduce (Metis)*

OS functionalities. After 1990 s, OS researchers gradually turn to the scalability problem, which put mosbench [12] on the stage as the first benchmark suite for OS scalability. Mosbench consists of seven application benchmarks including services, database, and offline analytics. We also find a microbenchmark-based suite, Will-It-Scale [4], that performing intensive invocations of system calls to see the system scalability.

2.2 What Benchmarks are OS Researchers Using?

To make it specific, we summary the benchmarks used by many OS research publications in Table 2. OS researchers use micro-benchmarks to stress particular subsystems for analysis of a single subsystem or component and application benchmarks to evaluate overall performance. Table 2 counts the number of micro and application benchmarks used by each system. Besides micro benchmarks mentioned above, many application benchmarks are used to gain insights or demonstrate effects.

Benchmarks in SPLASH-2 [36] and PARSEC [10] are kernel algorithms dervied from real world shared-memory programs. We classify them into application benchmarks as their OS behaviors are similar to in-production programs, even in a smaller scale. However, these benchmarks consumes only CPU and memory resources. For new designs, researchers persue other realistic applications to make the evaluation more convincing. For example, Corey [11] used Metis and webd as the application benchmarks for evaluating scalabitliy, and Barrelfish [8] chose httperf, lighttpd, and SQLite. However, researchers usually choose application benchmarks at will, because there is no benchmark suite designed for evaluating an OS with realistic big data applications!

We find that even on a machine with more than twenty cores, systems built on a monolithic kernel (like Corey) or as a single system image (like Barrelfish) were not evaluated using mixed workloads. This is unrelaistic for a multicore machine in production use. Luckily, some works did use mixed workloads, which are Celluar Disco, Xen, K42, Linux Containers, HeliOS, and Tessellation. For these works, we gain two observations: (1) Such sytems provide multiple execution envrionments in a single server. The system is designed for multiplex system resources among multiple applications, such as virtual machines, resource containers, satellite kernels, etc. (2) They choose mixed workloads according with instant wishes, which can be seen from the single simple benchmarking scenario for each system. Every evaluation includes only one mix of applications: (micro, micro), (application, application), or (application, micro). The system resources these benchmarks consume most cover all the system resources, even kernel resources (fork bomb used by Xen). Hardly can we find a persuasive reason for such combinations of benchmarks.

2.3 Why a Mixed Workload

There are several reasons that we need a mixed workload for effective evaluation of an operating system.

Table 2. Benchmarks used in some OS research papers in the past twenty years. (Benchmarks in *italic* are application benchmarks. Benchmarks in **bold** are used as mixed workloads).

Year	Systems	Properties	micro	app	Benchmarks
1995	Hive [15]	performance; fault containment	4	3	*SPLASH-2(RayTrace; ocean)*; *Pmake*; file read; file write; open file; page fault; fault injection (Raytrace or Pmake)
1997	Disco [14]	performance; scalability	0	4	*TPC-D on Informix Database*; *Pmake*; *SPLASH-2 (RayTrace)*; *SPEC WEB96 (Apache)*
1999	Tornado [20]	performance; scalability	7	0	Memory allocation; object miss handling; object garbage collection; procedure calling; thread Creation; in-core Page Fault; file stat
1999	Celluar Disco [21]	performamnce; isolation	0	4	*TPC-D (Informix)*; *parallel make (Pmake)*; *SPLASH-2(RayTrace)*; *SpecWEB96 (Apache)*; **TPC-D + RayTrace**
2003	Xen [6]	performance; scalability; isolation	5	2	SPEC INT2000; *OSDB (PostgreSQL)*; dbench(file system); *SPEC WEB99*; lmbench suite; **OSDB + SPEC WEB99 + dd + fork bomb**
2005	K42 [5,23]	scalability; isolation	0	2	*SPEC SDET*; **SPEC SDET + streaming applications**
2007	LinuxContainers [30]	performance; scalability; isolation	5	3	lmbench suite; iperf; dd; *dbench*; Postmark; CPU-intensive; *kernel compile*; *OSDB (PostgreSQL)*; **OSDB + dd**
2008	Corey [11]	scalability	5	2	memclone; mempass; a simple TCP service; object operations (global & local share); file duplication; *Metis*; *webd (filesum)*
2009	HeliOS [27]	performance; isolation	6	1	message passing(SingBench); netstack; PostMark; **SAT solver + a disk indexer**; scheduling stress test; *Mail server*
2009	Barrelfish [8]	scalability	7	3	message passing; NPB (CG, FT, IS); SPLASH-2 (Barnes-Hut, radiosity); ipbench; *httperf*; *lighthttpd*; *SQLite*
2010	fos [34,35]	performance; scalability	4	1	system call (local &remote); pings; process creation; file access; *ApacheBench*
2010	Linux [12](evaluation)	scalability	0	7	*Exim*; *memcached*; *Apache*; *PostgreSQL*; *gmake*; *Psearchy*; *MapReduce (Metis)*
2011	Cerburus [31]	performance	5	4	signal handling; process fork & clone; inter-VM message passing; file reading; network; *histogram*; *dbench*; *Apache*; *Memcached*
2012	Dune [9]	performance	11	3	getpid; page fault; page walk; ptrace; trace; appel1; appel2; SPEC2000; *Lighttpd*; *Wedge*; GCBench; Linked List; Hash Map; *XML parser*
2013	Tessellation [16]	performance; elasticity	1	2	NAS parallel benchmarks (EP); *a video player*; **video player + dropbox**
2014	K2 [25]	performance	4	0	DMA transfer; ext2fs accessing; UDP loopback; memory allocation

- In industry, workloads are widely consolidated on a virtualized system or even a monolithic operating system like Linux. It is an essential expectation that newly developed systems should meet industrial demands. In some scenarios, applicaitons are still deployed on independent servers because of the possible mutual interference brought by colocated workloads. As a result, when new OSes are implemented, we need to evaluate and report their tolerance to interference.

– Both performance isolation and scalability are affected by mixed workloads. Observations from running a single workload in terms of performance and scalability will be influenced by co-located workloads. For example, Ihor Kuz, etc. [24] compared the result of running a mixed workload with a single work-load. The performance curve of a single workload can be largely impacted if they corun another workload in the same server, indicating single-workload benchmarks alone are unlikely to provide sufficient insight into the working of the OS.

– Using tail latency [17] as the metric of both online services and kernel activi-ties, the performance curve with colocated workloads will deviate a lot from that without any interferer. Figures 1a and b respectively show the influences of an offline workload (streamcluster from PASREC 3.0 [10]) to the online workloads of memcached [1] and Search [33] and an offline workload (body-track from PASREC 3.0). We find that the perforamnce of online services can be more easily influenced by offline workloads. When coloating bodytrack streamcluster, the performance vairation is very limited in Linux Containers (LXC) and Xen. And the metric of througput and average latency can not let us fully understand the performance of Search colocated with other workloads. Tail latency is sensitive in the context of mixed workloads.

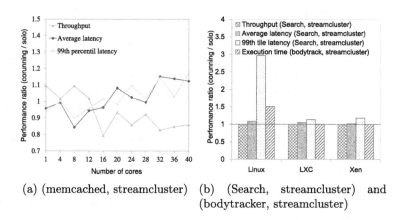

(a) (memcached, streamcluster) (b) (Search, streamcluster) and (bodytracker, streamcluster)

Fig. 1. Ratios of the performance of colocating workload A with B to the performance of solo running of workload A, which are denoted as (A, B).

3   Design of Mbench

Different from [24] which aims at finding an optimal mix that pushes the OS to its limits using micro benchmarks, we are building an operational, realistic, multi-dimensional benchmark suite for overall evaluation of an OS. We determine the properties an ideal benchmark suite should own as follows:

- It covers both micro and application benchmarks for an in-depth and wide evaluation. On the one hand, the collection of micro benchmarks should be able to stress individual subsystems of an OS, including resource management and kernel functionalities. Researchers desire sufficient insights into the inside of an OS. On the other hand, big data applications with realistic workloads should be included. It is high time to consider realistic big data applications that the industry calls attention to. Traditional application benchmarks selected by researchers as shown in Table 2 become obsolete with the fast changes in architectures [17], programming models [18,37], and 4 V of big data [33].
- Latency critical workloads are included. As mentioned in Sect. 2.3, latency critical applcations usually have physical demand on an isolated environment. Their performance is tightly tied to commerical interests. We should never ignore this category of workloads of which in-memory computing comes into the limelight. In particular, the phenomenon of the tail at scale [17] impels us to take into consideration the metric of tail latency.
- It should keeps as least redundant benchmarks as possible. If the benchmark suite consists of N benchmarks and we identify any i benchmarks as a candidate mix, the amount of mixes we need to evaluate will increase to be a combination: C_N^i. If a single round of execution of the benchmark suite needs T hours, the evaluation time of i-mix will be more than $C_N^* T/n$ hours. Redundant mixes exist in such naive combinations which fail to provide supplementary and useful information.
- A tool of managing experiments of mixed benchmarks can greatly boost the evaluation efficiency. Managing mixed benchmarks is an order of magnitude more difficult than just running a single application. Firstly, mixed benchmarks need to run at a coordinated pace, including the preparation phase and running phase. Secondly, logging and analysis of multiple benchmarks becomes complex since we may conduct a large number of experiments to fully understand the system. Thirdly, workload consolidation is usually conducted in virtualizied evnvironments like Linux containers and Xen. We should include such systems with multiple execution environments each of which is desinged for one benchmark. In these systems configuration of specified deployments are quite different, such as CPU affinity and memory allocation, so as the monitoring methods.

The rest of this section will address these concerns.

3.1   Benchmark Construction

The general pricinple is choosing benchmarks from existing publications and making sure the properties claimed above are acquired. The benchmarks we choose are listed in Table 3. All of them can be used to evaluate performance at a stand-alone mode, but our intention of putting them together is for performance isolation.

Micro Benchmarks: We choose the benchmarks stressing different resources in the system. Will-It-Scale [4] is a benchmark suites consisting 40+ test cases of intensively invoking different system calls. We modify it to support exporting latencies of each invocation. To reduce the evaluation efforts, we can further choose test cases from SPEC CPU or Will-It-Scale.

Application Benchmarks: There are some research work releasing real big data benchmarks with customizable workloads and data sets, such as BigData-Bench [33] and CloudSuite [19]. We choose Spark, Hadoop, and Search workloads from BigDataBench 3.1, memcached from mosbench [12] which is extended with request profiling, and PostgreSQL from the Open Source Database Benchmark [2]. Multicore applications in the server can be classified into three categories: traditional offline batches that independently accomplish a short task, online services that serve requests from clients, and big data analytics programs built on new programming models. Offline batches and analytics identify execution time as the metric, while online services identify throughput and latency (in particular tail latency) as their metrics. Meanwhile, the resources utilized by these applications can be profiled according to single-application experiments, which are listed in priority order in Table 3. We try to coverage both the dimention of application category and the dimention of resource utilization in the benchmark selection. As a result, the redundancy can be highly limited.

3.2   Usage of Mbench

We take two problems cared by OS researchers for examples to explain the usage of Mbench.

First, how are the services concurrent with other applications? This problem makes an asummption that the performance of services is critical. That is, we should consider a service application as the foreground workload and other applications as background interferers. In this scenario, we settle the service application and change the interfering workloads from consuming one resource to another. Before adding interferers, we need to find the best performance of one workload, and then gradually increase the pressure stressed by the background workloads. Note that the metric of tail latency is of more information in a large scale cluster. Moreover, two or more concurrent interferers can be also investigated.

Second, what workload mixes yield the most information when used to evaluate a multicore OS? To put the system to the edge of performance, we have two methods: one is that combining benchmarks contending for the same resources; the other is that using the method of linear programming presented in [24]. For the first method, the microbenchmarks listed in Table 3 are enough. Running multiple instances of a microbenchmark can probably make the OS expose the

Table 3. The summary of benchmarks included in Mbench

Type	Benchmark	Resource priorities	Workload type	Metrics
Micro benchmarks	SPEC CPU (bzip2, sphix3)	CPU	-	execution time
	cachebench	cache, memory	-	bandwidth
	IOzone	filesystem (disk)	-	bandwidth
	netperf	network (ethernet)	-	bandwidth
	Will-It-Scale [4]	kernel functionalities	-	throughput, tail latency
Application benchmarks	PARSEC [10] (bodytrack, streamcluster)	CPU, memory, file system	offline batch	execution time
	memcached	memory, CPU, network	service	throughput, tail latency
	Spark (kmeans, pagerank)	memory, CPU, network	analytics	execution time
	PostgreSQL	file system, memory, CPU	service	throughput, tail latency
	Hadoop (sort, grep)	file system, memory, CPU	analytics	time
	Search (tomcat; nutch)	network, memory, CPU; CPU, memory, network	service	throughput, tail latency

bottlenecks, which does not require many experimentation efforts. And Will-It-Scale will help further locate the problems as it consists test cases of most kernel functionalities. For the latter method, we need to obtain the matrix of resource usage and performance for all benchmarks. This deserves much experimentation efforts but can prabably lead to an optimal solution through the integer linar programming.

In a nutshell, we can design the mixes according to the resources consumed by the benchmarks to reduce the efforts of performing experiments and focus on specified areas.

3.3   Experiment Control

To faciliate the supervising of mixed workloads, we develop an experiment controlling tool, *Mcontroller*. It adopts an object oriented programming paradigm, written in Python, to simplify the relationships between experiments, tests, and applications. An experiment may include many tests and a test may include many applications. Deployments and executions are enforced according to formalized configurations of an expeirment, a test, or an application. We note that large companies has recently published their work on large-scale cluster management [28,32], whose focuses are task scheduling and resource adjustment. Our intention is not to compete with them in cluster mangement, but to present a simplified management of statically deployed applications targeting at different OSes. Naturally, we can extend the tool with functions of resource adjustment

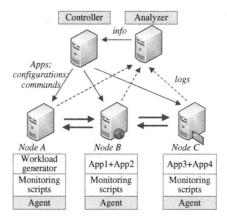

Fig. 2. Structure of the experiment control tool (Mcontroller).

or task scheduling for purposes of evaluation of the system elasicity or a certain scheduling algorithm. Current design of Mcontroller is modular and capable of containing such extensions.

Figure 2 depicts the general structure of Mcontroller. It mainly consits of three components: *Controller*, *Analyzer*, and *Agent*. Controller reads configurations, replicates the benchmarks, distributes an Agent on each host (or an execution environment), and controls the lifetime of the benchmarks through issuing commands to the Agents. Agent forks processes to carry out the commands from Controller or Analyzer. The inner communication subsystem is a RPC mechanism from mosbench [12]. An Agent exhibits little overhead in network transmission and process mangement, which can be ignored for the running of benchmarks. Analyzer collects logs from specified agents and provide interfaces of data analysis. We add supports of online monitoring and analysis of tail latency performance.

Current implementation supports applications mentioned above and systems including Linux, Linux containers, and Xen. The monitor scripts can collect OS-level (through the *proc* subsystem) and architecture-level (through both perf and Oprofile, XenOprofile for Xen) characteristics. Moreover, we also provide templates for extentions of more benchmarks. We are continuing to improve the avaiability and benchmark coverage, hoping it can provide more facilities.

4   Use Cases

Co-running online services and offline batches is always a thorny problem. We try to investigate performance isolation of Linux, LXC, Xen in terms of co-running Search with batch workloads. In this test, each OS instance runs on 6 cores and 16 GB memory within a NUMA node. We run a Nutch back end and PARSEC workloads respectively on the two OS instances (container or virtual machine). The requests are replayed in a uniform distribution at 300 requests/s beyond

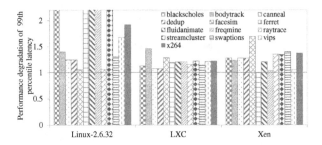

Fig. 3. Tail latency degration of Search when co-locating with Parsec benchmarks relative to running Search alone. The baseline tail latencies when running Nutch alone on Linux, LXC, and Xen are 108.6, 129.3, and 210.9 ms, respectively.

which the tail latency of XEN will be degraded dramatically. Figure 3 illustrates performance degradation of tail latency compared to the baselines of running a Nutch back end alone on one OS instance of each system. Xen has the poorest tail latency performance (210.9 ms) when running a single Nutch server at 300 req/s. Actually, even at 250 req/s, its tail latency is still as high as 150.6 ms, greater than Linux and LXC, and the average degradation reaches 25.6 %. As virtualization overhead exists, Xen is not suited for a high load level, and it behaves worse than LXC in the figure. Actually, Xen has better performance isolation than Linux in terms of average performance of offline batches [6]. But we can see that for latency critical services the tail latency is much deteriorated, because co-located guest OSes share resources provided by VMM and particularly can be interfered by VMM activities and competitions from others.

5   Discussion and Conclusion

We claim that existing multicore OS benchmarks can not reveal information of OS benhaviors when running several applications concurrently in a single server. Increasing density of hardware resources makes workload consolidation much popular in production data centers, which calls urgent attentions for evaluating new designs and technologies in the OS literature using mixed, realistic, and dynamic workloads. We propose a new benchmark suite to provision mixed workloads with not only traditionaly microbenchmarks, but also real world big data and latency critical applications. We develop an distributed extensible tool for experiment controlling and result analysis. We are studying use cases of leveraging the benchmarks to evaluating existing operating systems (Linux, LXC, and Xen) and paln to release Mbench in few months.

In the future, we intend to further analyze the OS-level characteristics of the benchmarks. Choosing a good mix is different from choosing a minimum colleciton covering all the characteristics. As hardware resources of a server vary with the time, we should consider mixes of more workloads rather than current two or three to obtain higher efficiency. Portability is also a thorny problem for some new OSes designed with changed APIs. It is our intention to work with others from the production data centers to further track their needs of consolidating big data applications, and to develop a more general framework for producing benchmark suites used for future OS researchs in particular operating systems designed for heterogeous or manycore platforms.

Acknowledgements. This work is supported by the National High Technology Research and Development Program of China (Grant No. 2015AA015308), the Major Program of National Natural Science Foundation of China (Grant No. 61432006), and the Key Technology Research and Development Programs of Guangdong Province, China (Grant No.2015B010108006).

References

1. Memcached. http://memcached.org/. Accessed December 2014
2. The open source database benchmark. http://osdb.sourceforge.net/. Accessed June 2015
3. Unixbench. https://github.com/kdlucas/byte-unixbench. Accessed June 2015
4. Will-it-scale benchmark suite. https://github.com/antonblanchard/will-it-scale. Accessed June 2015
5. Appavoo, J., Auslander, M., Butrico, M., da Silva, D.M., Krieger, O., Mergen, M.F., Ostrowski, M., Rosenburg, B., Wisniewski, R.W., Xenidis, J.: Experience with K42, an open-source, linux-compatible, scalable operating-system kernel. IBM Syst. J. **44**(2), 427–440 (2005). http://dx.doi.org/10.1147/sj.442.0427
6. Barham, P., Dragovic, B., Fraser, K., Hand, S., Harris, T., Ho, A., Neugebauer, R., Pratt, I., Warfield, A.: Xen and the art of virtualization. ACM SIGOPS Operating Syst. Rev. **37**(5), 164–177 (2003)
7. Barroso, L.A., Clidaras, J., Hölzle, U.: The Datacenter as a Computer: An Introduction to the Design of Warehouse-Scale Machines. Synthesis Lectures on Computer Architecture. Morgan and Claypool, San Rafael (2013)
8. Baumann, A., Barham, P., Dagand, P.E., Harris, T., Isaacs, R., Peter, S., Roscoe, T., Schüpbach, A., Singhania, A.: The multikernel: a new OS architecture for scalable multicore systems. In: Proceedings of the ACM SIGOPS 22nd Symposium on Operating Systems Principles, pp. 29–44. ACM (2009)
9. Belay, A., Bittau, A., Mashtizadeh, A., Terei, D., Mazières, D., Kozyrakis, C.: Dune: Safe user-level access to privileged CPU features. In: Proceedings of the 10th USENIX Conference on Operating Systems Design and Implementation OSDI 2012, pp. 335–348. USENIX Association (2012)
10. Bienia, C., Kumar, S., Singh, J.P., Li, K.: The PARSEC benchmark suite: characterization and architectural implications. In: Proceedings of the 17th International Conference on Parallel Architectures and Compilation Techniques, pp. 72–81. ACM (2008)
11. Boyd-Wickizer, S., Chen, H., Chen, R., Mao, Y., Kaashoek, F., Morris, R., Pesterev, A., Stein, L., Wu, M., Dai, Y., Zhang, Y., Zhang, Z.: Corey: An operating system for many cores. In: Proceedings of the 8th USENIX Conferenceon Operating Systems Design and Implementation OSDI 2008, pp. 43–57. USENIX Association, Berkeley (2008). http://dl.acm.org/citation.cfm?id=1855741.1855745

12. Boyd-Wickizer, S., Clements, A.T., Mao, Y., Pesterev, A., Kaashoek, M.F., Morris, R., Zeldovich, N.: An analysis of linux scalability to many cores. In: Proceedings of the 9th USENIX Conference on Operating Systems Design and Implementation OSDI 2010, pp. 1–8. USENIX Association, Berkeley (2010)
13. Brown, A.B., Seltzer, M.I.: Operating system benchmarking in the wake of lmbench: A case study of the performance of NetBSD on the intel x86 architecture. In: Proceedings of the 1997 ACM SIGMETRICS International Conference on Measurement and Modeling of Computer Systems SIGMETRICS 1997, pp. 214–224. ACM, New York (1997)
14. Bugnion, E., Devine, S., Govil, K., Rosenblum, M.: Disco: Running commodity operating systems on scalable multiprocessors. ACM Trans. Comput. Syst. (TOCS) 15(4), 412–447 (1997)
15. Chapin, J., Rosenblum, M., Devine, S., Lahiri, T., Teodosiu, D., Gupta, A.: Hive: Fault containment for shared-memory multiprocessors. In: Proceedings of the Fifteenth ACM Symposium on Operating Systems Principles SOSP 1995, pp. 12–25 (1995)
16. Colmenares, J.A., Eads, G., Hofmeyr, S., Bird, S., Moretó, M., Chou, D., Gluzman, B., Roman, E., Bartolini, D.B., Mor, N., Asanović, K., Kubiatowicz, J.D.: Tessellation: Refactoring the OS around explicit resource containers with continuous adaptation. In: Proceedings of the 50th Annual Design Automation Conference DAC 2013 (2013)
17. Dean, J., Barroso, L.A.: The tail at scale. Commun. ACM 56(2), 74–80 (2013)
18. Dean, J., Ghemawat, S.: Mapreduce: simplified data processing on large clusters. Commun. ACM 51(1), 107–113 (2008)
19. Ferdman, M., Adileh, A., Kocberber, O., Volos, S., Alisafaee, M., Jevdjic, D., Kaynak, C., Popescu, A.D., Ailamaki, A., Falsafi, B.: Clearing the clouds: a study of emerging scale-out workloads on modern hardware. In: Proceedings of the Seventeenth International Conference on Architectural Support for Programming Languages and Operating Systems ASPLOS 2012, pp. 37–48 (2012)
20. Gamsa, B., Krieger, O., Appavoo, J., Stumm, M.: Tornado: Maximizing locality and concurrency in a shared memory multiprocessor operating system. In: Proceedings of the Third Symposium on Operating Systems Design and Implementation OSDI 1999, pp. 87–100. USENIX Association, Berkeley (1999). http://dl.acm.org/citation.cfm?id=296806.296814
21. Govil, K., Teodosiu, D., Huang, Y., Rosenblum, M.: Cellular disco: Resource management using virtual clusters on shared-memory multiprocessors. In: ACM SIGOPS Operating Systems Review, vol. 33, pp. 154–169. ACM (1999)
22. Juckeland, G., Kluge, M., Nagel, W.E., Pfluger, S.: Performance analysis with BenchIT: portable, flexible, easy to use. In: International Conference on Quantitative Evaluation of Systems, pp. 320–321 (2004)
23. Krieger, O., Auslander, M., Rosenburg, B., Wisniewski, R.W., Xenidis, J., Da Silva, D., Ostrowski, M., Appavoo, J., Butrico, M., Mergen, M., et al.: K42: building a complete operating system. In: ACM SIGOPS Operating Systems Review, vol. 40, pp. 133–145. ACM (2006)
24. Kuz, I., Anderson, Z., Shinde, P., Roscoe, T.: Multicore OS benchmarks: We can do better. In: Proceedings of the 13th USENIX Conference on Hot Topics in Operating Systems HotOS 2013, pp. 10–10. USENIX Association, Berkeley (2011). http://dl.acm.org/citation.cfm?id=1991596.1991610
25. Lin, F.X., Wang, Z., Zhong, L.: K2: A mobile operating system for heterogeneous coherence domains. In: Proceedings of the 19th International Conference on Architectural Support for Programming Languages and Operating Systems, pp. 285–300. ACM (2014)

26. McVoy, L., Staelin, C.: Lmbench: Portable tools for performance analysis. In: Proceedings of the 1996 Annual Conference on USENIX Annual Technical Conference ATEC 1996, p. 23. USENIX Association, Berkeley (1996)
27. Nightingale, E.B., Hodson, O., McIlroy, R., Hawblitzel, C., Hunt, G.: Helios: heterogeneous multiprocessing with satellite kernels. In: Proceedings of the ACM SIGOPS 22nd Symposium on Operating Systems Principles, pp. 221–234. ACM (2009)
28. Schwarzkopf, M., Konwinski, A., Abd-El-Malek, M., Wilkes, J.: Omega: Flexible, scalable schedulers for large compute clusters. In: Proceedings of the 8th ACM European Conference on Computer Systems EuroSys 2013, pp. 351–364. ACM, New York (2013). http://doi.acm.org/10.1145/2465351.2465386
29. Schwarzkopf, M., Murray, D.G., Hand, S.: The seven deadly sins of cloud computing research. In: Proceedings of the 4th USENIX Conference on HotTopics in Cloud Computing HotCloud 2012, pp. 1–1. USENIX Association, Berkeley (2012). http://dl.acm.org/citation.cfm?id=2342763.2342764
30. Soltesz, S., Pötzl, H., Fiuczynski, M.E., Bavier, A., Peterson, L.: Container-based operating system virtualization: a scalable, high-performance alternative to hypervisors. In: ACM SIGOPS Operating Systems Review, vol. 41, pp. 275–287. ACM (2007)
31. Song, X., Chen, H., Chen, R., Wang, Y., Zang, B.: A case for scaling applications to many-core with OS clustering. In: Proceedings of the Sixth Conference on Computer Systems EuroSys 2011, pp. 61–76. ACM, New York (2011). http://doi.acm.org/10.1145/1966445.1966452
32. Verma, A., Pedrosa, L., Korupolu, M., Oppenheimer, D., Tune, E., Wilkes, J.: Large-scale cluster management at google with borg. In: Proceedings of theTenth European Conference on Computer Systems EuroSys 2015, pp. 18:1–18:17. ACM, New York (2015). http://doi.acm.org/10.1145/2741948.2741964
33. Wang, L., Zhan, J., Luo, C., Zhu, Y., Yang, Q., He, Y., Gao, W., Jia, Z., Shi, Y., Zhang, S., Zheng, C., Lu, G., Zhan, K., Li, X., Qiu, B.: Bigdatabench: A big data benchmark suite from internet services. In: 2014 IEEE 20th International Symposium on High Performance Computer Architecture (HPCA), pp. 488–499, February 2014
34. Wentzlaff, D., Agarwal, A.: Factored operating systems (fos): the case for a scalable operating system for multicores. ACM SIGOPS Operating Syst. Rev. 43(2), 76–85 (2009)
35. Wentzlaff, D., Gruenwald, III, C., Beckmann, N., Modzelewski, K., Belay, A., Youseff, L., Miller, J., Agarwal, A.: An operating system for multicore and clouds: Mechanisms and implementation. In: Proceedings of the 1st ACMSymposium on Cloud Computing SoCC 2010, pp. 3–14. ACM, New York (2010). http://doi.acm.org/10.1145/1807128.1807132
36. Woo, S., Ohara, M., Torrie, E., Singh, J., Gupta, A.: The SPLASH-2 programs: characterization and methodological considerations. In: 1995 Proceedings of the 22nd Annual International Symposium on Computer Architecture, pp. 24–36, June 1995
37. Zaharia, M., Chowdhury, M., Das, T., Dave, A., Ma, J., McCauley, M., Franklin, M.J., Shenker, S., Stoica, I.: Resilient distributed datasets: A fault-tolerant abstraction for in-memory cluster computing. In: Proceedings of the 9th USENIX Conference on Networked Systems Design and Implementation, p. 2. USENIX Association (2012)

Performance Optimization
and Evaluation

Evolution from Shark to Spark SQL: Preliminary Analysis and Qualitative Evaluation

Xinhui Tian[1,2(✉)], Gang Lu[1,2], Xiexuan Zhou[1,2], and Jingwei Li[3]

[1] State Key Laboratory of Computer Architecture,
Institute of Computing Technology, Chinese Academy of Sciences, Beijing, China
`tianxinhui@ict.ac.cn`
[2] University of Chinese Academy of Sciences, Beijing, China
[3] Beijing Academy of Frontier Science and Technology, Beijing, China

Abstract. Spark is a general distributed framework with the abstraction called resilient distributed datasets (RDD). Database analysis is one of the main kinds of workloads supported on Spark. The SQL component on Spark has evolved from Shark to Spark SQL, while the core components of Spark also have evolved a lot comparing with the original version. We analyzed on which aspects Spark have made efforts to support many workloads efficiently and whether the changes make the support for SQL achieve better performance.

1 Introduction

The requirements of big data analytics has grown rapidly in recent years, which facilitates the development of various large scale data processing systems, such as MapReduce [4], Dryad [7] and Dremel [11]. Among these systems, Spark [17] has attracted a lot of attention in the last few years, with the conception of resilient distributed datasets (RDD) and good performance.

Like many other large scale data processing systems, Spark is an open-source framework, and has attracted more than 400 developers to contribute, which makes the code bases increase rapidly. Spark has now evolved from a specific computing platform to a much more general big data processing system with the support of various types of applications, such as database query, graph processing, and streaming computing. Moreover, a lot of new features have been added for more efficient memory management and better reliability.

However, the rapid development of Spark also brings some problems for both users and developers. For example, the continuous increasing configurations make it very hard for performance tuning of different workloads. Moreover, as Spark adopts a general distributed engine to support different types of workloads, improvements on the core components can significantly impact the execution procedure and performance of some specific types of workloads with the configurations unchanged.

We suggest that a comprehensive study on the evolution of Spark is very important in this scenario. By studying the evolution of Spark with different versions, the following questions can be answered: What components of Spark

© Springer International Publishing Switzerland 2016
J. Zhan et al. (Eds.): BPOE 2015, LNCS 9495, pp. 67–80, 2016.
DOI: 10.1007/978-3-319-29006-5_6

have changed a lot and what the reasons are for the changes? Which performance and reliability features are added? What components are related to each configuration? What features are added to Spark to support specific workload types? Answers of these questions can be very helpful for both users who want to tune their Spark applications and researchers who are trying to design a good benchmark for Spark system. The observations from our work can also help developers who want to improve current Spark system, or create new big data systems.

In this paper, we present our preliminary work of this study, in which we analyze the evolution of the database query processing framework in Spark. The database part is chosen for two reasons. First, database query processing has long been one of the most important types of workloads Spark community caring about, which has the most issues on Spark JIRA [2]. Second, the hierarchy has been changed a lot and the query compiler has been almost rewritten. Database query was first supported on Spark via introducing a parse layer called Shark [15], which is inspired by Hive [13] and translates a SQL query into a Spark job. Since version 1.0.0, a new component called Spark SQL [3] is introduced to replace Shark, which provides the ability to load data from existing RDDs and use procedural APIs of other libraries easily. Detailed Comparisons are presented on two layers: the Spark core layer and the query compiler. We also perform qualitative evaluation on these two platforms, and the results showed that Spark SQL could achieved better performance when the memory was big enough, and Shark had better plan optimizations.

We choose two versions of Spark to study in this paper, which are Shark 0.9.1 with Spark 0.9.1 and Spark 1.3.1 with Spark SQL respectively. In Sect. 2, we first give a brief overview of the Spark system and the two SQL compilers. An analysis on the changes of Spark core components is presented in Sect. 3. In Sect. 4, we demonstrate the differences between Shark and Spark SQL when they phase SQL queries. We will then give the experiment results in Sect. 5, related works in Sect. 6, and conclusions in Sect. 7.

2   An Overview of Spark, Shark, and Spark SQL

In this section, we first give a brief description on Spark framework, and then introduce Shark and Spark SQL.

2.1   Spark

Spark is a cluster computing framework based on the abstraction of resilient distributed dataset (RDD). An RDD represents a read-only collection of objects partitioned across a set of machines, which can be rebuilt when a partition is lost. Spark provides the interface for users to decide whether to cache an RDD across machines. This feature makes it easy to reuse data which makes Spark well-suited for iterative jobs. For fault tolerance, each RDD records the transformations used to build it. The RDD then contains enough information about how it is derived from other RDDs, thus if a partition is lost, this RDD can recompute just that partition to recover the lost data.

RDD have two types of operations: transformation and action. The transformation operations, such as map, filter, and join, represent how an RDD is created from data in stable storage or other RDDs. The action operations are the operations on one RDD and return a value, such as collect, count, and save. As mentioned above, the transformations on each RDD is stored as a set of dependencies on parent RDDs. The dependencies are classified into two types: narrow dependencies where each partition of a parent RDD is used by at most one partition of the child RDD, and wide dependencies where multiple child partitions depend on one parent partition.

For job execution, Spark adopts a delay scheduling mechanism, which means a job will not be submitted to the driver unless an action operator is called. This mechanism enables Spark to perform optimizations before a job starts to execute. When a job is submitted, it is transferred from a RDD's lineage graph to a directed acyclic graph (DAG) of stages for execution. Wide dependencies (i.e., shuffle dependencies) with shuffle operations are spilt into different stages, while the narrow dependencies are computed in a pipelined way in one stage. The scheduler then launch tasks of each stage on each machine to compute the missing partitions until the target RDD has been computed. The task scheduler assigns tasks based on data locality, which means to send each task to the node containing the data partition. Results of each job are finally sent back to the driver or stored in a storage system according to the action operation.

2.2   Shark and Spark SQL

Shark is a query processing layer built on the Spark engine. Shark provides in-memory columnar storage and columnar compression to store and process relational data efficiently. For query processing, Shark uses a three-step similar as traditional RDBMS, which includes query parsing, logical plan generation and physical plan generation. Shark uses Hive query compiler directly for query parsing and generates an abstract syntax tree (AST) for each query. The AST is then translated into a logical plan consisting of operators and is later optimized through some basic rules. Shark compiles the logical plan to create a physical plan composed of transformations on RDDs. The physical plan is then submitted to the Spark master as a Spark application and is executed by the engine.

Even though Shark provides the ability of SQL query processing on Spark, there are still some limitations of it. For example, the logical plan generated by Hive is originally designed for Hadoop MapReduce, which is not very suitable for the Spark engine. Results computed from Shark also can not be used by applications of other types directly, which makes it difficult to perform more complex analytics. Spark SQL is developed as a module of Spark framework. Spark SQL introduces a new abstraction called DataFrame to support the wide range of data sources and algorithms. A DataFrame is a distributed collection similar to a table in a relational database, with data from either external sources or Spark's built-in distributed collections. Users can then manipulate records in this DataFrame using Spark's procedural API or relational APIs provided by Spark SQL. Spark SQL also supports nested data model from Hive for tables for Hive sql compatibility.

For query processing, Spark SQL uses a new component called Catalyst instead of the Hive compiler for logical plan generating and optimizing. Catalyst is an extensible optimizer which provides the convenience for external developers to add new optimization rules and features. Similar to Shark, a query in Spark SQL is first translated into an AST, and then into a logical plan. The logical plan is then optimized by Catalyst with some rule-based optimizations, such as constant folding, predicate pushdown. Spark SQL generates one or more physical plans for one optimized logical plan, and uses a cost model to select one for processing. Spark SQL also adopts code generation to optimize the final plan for better performance.

3   The Evolution of Spark Core Components

In this section, we present some main changes of Spark core components from version 0.9.1 to 1.3.1. The components we analyzed include the scheduler (DAG scheduler and task scheduler), the shuffle components (shuffle mechanism, shuffle write and shuffle read), the communication components (block and message transfer service), the storage components (block manager, cache manager and other storage layers support), and the memory management mechanisms (mainly garbage collection).

We do not list other components here (such as the RDD abstraction, the executor backend, and the broadcast components). One reason is that these components don't have some significant changes during the evolution, or the changes are mainly bug fixes. We list the main changes we found in Table 1. Most of the changes are collected from the patches and change logs released from Spark communities. We also find some other interesting differences by comparing source codes of those two versions. For example, the event processing mechanism of DAG scheduler has changed from actor to an event processing thread, but the released note did not mention it.

Through Table 1, we can get some conclusions about which aspects have gained lots of attention and the main points of optimization: (1) The structure of Spark source code has been modified to be more modularized and more flexible for developers to add new mechanisms for different components. Many components, such as shuffle, storage, and communication service, have added a manager class to provide public interfaces for extension. (2) Lots of optimizations and improvements aim to improve the performance of Shuffle and communication, including serializing the task structures to reduce the size of data for broadcast, adopting Netty for lower latency, and introducing sort based shuffle mechanism for better scalability and performance. (3) Improvements on storage components mainly focus on reducing the usage of memory and memory more efficient management, since the memory resource are very precious in Spark. A new trend in Spark is to use third party systems for memory storage management. For example, Spark has added the support of Tachyon, which is an in-memory storage system towards high throughput writes and reads. (4) When performing computations, Spark relies mainly on JVM for memory management. Therefore the performance of garbage collection (GC), which is responsible for collecting memory of objects that are not used any more by the applications, is very essential in Spark. Spark has added many services to try to promptly release the unused states for GC. (5) The wide use of serialization is also an important trend in Spark. Spark developers try to use serialization to reduce the size of data for transmission. However, the CPU utilization can increase significantly as the result of such improvement, which can become a potential performance bottleneck.

Table 1. Changes between 0.9.1 and 1.3.1

		0.9.1	1.3.1
Scheduler	DAG Scheduler Event Processing	An event processing actor	An independent event processing thread
	Shared Task Structures Serialization	Not serializing the shared rdd and dep structures	Serializing the rdd and dep to reduce size of shared structures
	Task Description Serialization	Not serializing before sending to executor	Serializing the description before sending
Shuffle	Mechanism	Hash-based shuffle	Adding sort-based shuffle and setting it as default
	Shuffle Manager	No specific manager for shuffle	Adding shuffle manager to choose block manager for different mechanisms
	Shuffle Block Manager	Belonging to storage management, only supporting file block manager	Separating from storage component, supporting file and index shuffle block manager (designed for sort-based shuffle)
Communication	Block Transfer	Java nio as default	Adding block transfer service manager, and using Netty as default
	Communication Manager for Block Transfer	Nio-based	Netty-based
Memory Management	Garbage Collection	Using a metadataCleaner class to periodically clean up states	Using a daemon thread ContextCleaner and weak reference to clean up states
Storage	Cache Manager	First trying to put data into memory, if cannot hold, dropping to disk	Directly putting data into disk if the size is large
	Disk Store	Always using memory map for data reading	If the size of file is small, directly reading without using memory map
	Tachyon Support	No	Yes, supporting off-heap storage in Tachyon

4   Query Analysis

The queries we use in this paper are from AMPLab Big Data Benchmark [1]. Two tables are used in this queries, which are a Ranking table recording pagerank information of a web page, and a UserVisits table containing user visit records. Queries include one simple selection, one aggregation, and one join query. We will first analyze how Shark and Spark SQL parse and optimize the queries in details, then give a summary on the differences.

Selection. The selection query uses a where condition to select data from the Ranking table. For both platforms, a table read operator is first executed to get data from table. The filter condition has already been pushed down by the optimizer of Hive and Spark SQL, which can avoid unnecessary columns reading. All the operations can be executed in parallel in one stage. The difference is that Shark has to write the selected data into HDFS first and then fetch the results, since it uses the plan generated from Hive compiler. Hive sets the FileSinkOperator as the final operator for every query, with the consideration of reliability. Spark SQL can directly send the results to the shell, so it can achieve lower latency and better performance for such kind of queries.

Aggregation. The aggregation query performs two aggregation operations, substring and summary, and a group by operation on the UserVisits table. Shark splits the GroupByOperator generated from Hive into two sub operators called GroupByPreShuffleOperator and GroupByPostShuffleOperator. The GroupByPreShuffleOperator performs map side aggregations, where partial aggregations are executed for each key set in group by. A ReduceSinkOperator is then executed to repartition the results of map side aggregation for shuffle. The intermediate data are materialized to disks in a serialized key-value format. We need to mention that ReduceSinkOperators are used to break a query job into stages in Shark. The GroupByPostShuffleOperator is responsible for the final aggregation, which performs the aggregation operations for each key. The required columns are then selected and written into the HDFS finally.

Spark SQL uses an exchange and an aggregate operator to replace the ReduceSinkOperator and GroupByOperator respectively. The group by and aggregation operations are included in the aggregate operator, while the exchange operator is responsible for data repartition. The aggregation is also split into two phases, which are partial map side aggregation and a final aggregation. An exchange operator is the boundary of two stages which is similar as ReduceSinkOperator. The difference is that, the intermediate data are written to the disks with the same format as other rdds, instead of the special key-value format in Shark.

Join. The join query is much more complex, which contains a join operation of two tables, a group by operation, an order by operation and two aggregations with a limit condition. We give the final execution plans of two platforms in Fig. 1 to show the differences clearly. Shark and Spark SQL both start with table scan operations of two tables to get the required data and repartition the outputs using the join key for hash. In the next stage, a join operation is then executed to join the two datasets based on the join key. Aggregations are executed in the same way as mentioned above to compute the summary of adRevenue from UserVisits as totalRevenue, and the average value of pageRank

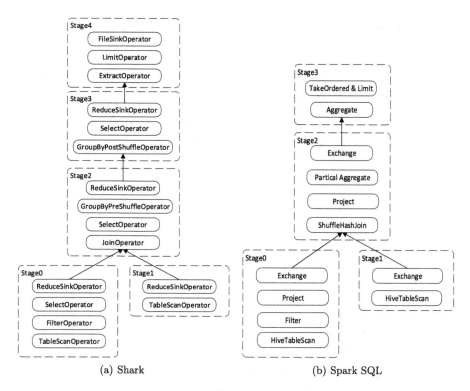

(a) Shark   (b) Spark SQL

Fig. 1. Execution plan of join query of Shark and Spark SQL

from Rankings table. The final operation is to pick the rows with the largest totalRevenue value (the order by operation and the limit condition). Since there is a ReduceSinkOperator here for Shark, the job is split into two stages, which first repartition the results according to totalRevenue, then get the rows with the largest values. In Spark SQL, this operation is much simpler. It uses a reduce operator of the rdd structure, which gets the rows with largest totalRevenue from each partition, and performs a comparison on the driver side to select the required rows.

Summary. From the above analysis, we can see that Shark indeed has some limitation for optimization by directly adopting the Hive compiler. For example, data need to be repartitioned and materialized whenever a ReduceSinkOperator is executed. As Shark uses the ReduceSinkOperator as the boundaries of stages, it can introduce unnecessary disk overhead and remain little space for optimizations based on different operations, such as the order by operation in the join query. The query compiler of Spark SQL is more flexible, and can generate an execution plan more suitable for the rdd abstraction. However, we also observed that Shark serializes the intermediate data before writing to disks, which is not found in Spark SQL. Such a operation can make the size of intermediate data smaller even though it costs more time to deserialize for later computation.

5   Experiment and Discussion

In this section, we perform experiments to evaluate the performance of Shark and Spark SQL with three queries. The cluster we use contains nine machines, one as master and other eight as workers. Each node has a Intel E5645 CPU with 6 cores, 32 GB memory, 1T HDD, and two NetXtreme II BCM5709 Gigabit Ethernet network adaptors. For Spark, we set 6 cores and 25 GB for each worker. More detailed configurations are shown in Table 2.

Table 2. Configuration of Spark

spark.shuffle.manager	hash
spark.shuffle.compress	true
spark.shuffle.memoryFraction	0.2
spark.storage.memoryFraction	0.2
spark.default.parallelism	300
spark.serializer	JavaSerializer
spark.shuffle.consolidateFiles	false

We used HiBench [6] to generate 4.9 GB data for Rankings table and 105.2 GB for UserVisits table. These two tables were stored in HDFS using RCFile format for good column scan performance. We ran each query with three different parameters varying selectivity to evaluate performance on different scale of data (We use numbers 1, 2, 3 to stand for different parameter, queries such as selection with the first parameter is represented as selection 1, etc. The data size selected increases as the number changes from 1 to 3.). We ran each query five times and computed the average processing time. The processing time information is shown in Fig. 2. We have the following observations: (1) Spark SQL performed better than Shark on all the select queries, which only process table scanning operations. (2) Spark SQL also achieved better performance when the size of data is not very large on aggregation and join queries. (3) Spark SQL with

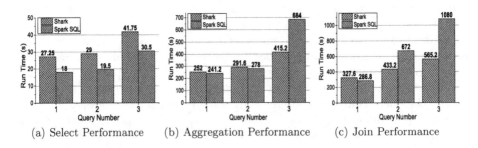

(a) Select Performance   (b) Aggregation Performance   (c) Join Performance

Fig. 2. Query performance of Spark SQL and Spark

Spark 1.3.1 began to perform worse than Shark with Spark 0.9.1 when the data size increased to a large scale (The data size selected from Uservisits table in join query 3 is 11.6 GB).

The good performance on table scanning of Spark SQL can be explained by removing the file sink operations during query execution. Improvements on communication mechanism and memory management can be the reasons why Spark SQL with Spark 1.3.1 can achieve better performance on aggregation 1, 2 and join 1.

To find the factors causing performance decrease in Spark SQL when selected data increased, we continued to collect low level information including CPU utilization and GC time from one worker during the execution of aggregation 3 and join 3 query (two queries with the largest data size selected, and Spark SQL achieved much poorer performance than Shark during the execution of these two queries). The results are shown in Figs. 3 and 4, integrated with the start time of every stage.

From Fig. 3(a) and (b), we can see that the CPU utilization of Spark SQL is much higher than Shark in stage 1 for aggregation 3 (the reason why the utilization is almost zero after 300 s is that this worker was waiting for the tasks on other workers to finish), we believe this is because Spark 1.3.1 performs many serialization and deserialization operations, however, the high CPU utilization does not help Spark 1.3.1 to achieve better performance. Both Spark SQL and Shark achieved a high CPU utilization when performing the join operation during join 3 query.

We got some interesting results when we compared the GC time distribution on these two platforms. As shown in Fig. 4, the GC time of Spark SQL is much longer than that of Shark, when join and aggregation operations are executed. That happens during the shuffle fetch phase when result tasks try to fetch the intermediate data, and long GC time means that the memory is not enough for tasks to execute computations.

Discussion. After analyzing the job logs of Spark SQL and Shark, we find that the main reasons for the pool performance of Spark SQL on Spark 1.3.1 are on two aspects: the shuffle mechanism on Spark core and query optimization rules on query compiler layer. For shuffle operations, the size of intermediate data Spark SQL generated is also much larger than that in Shark during shuffle writes. Moreover, for the shuffle read operations, Spark SQL on Spark 1.3.1 had to fetch more data that Shark on Spark 0.9.1 (18.4 GB vs 11.9 GB for aggregation 3, and 23.4 GB vs 9.6 GB for join 3 when performing join operation). For example, after scanning the 4.9 GB Rankings table, Spark SQL generated 7.9 GB data, while Shark still get 4.8 GB. That may be caused by the differences on intermediate data format and serialization we mentioned in Sect. 4. For query optimization, SQL operations, such as join and aggregation, are optimized better in Shark to fetch less data during shuffle reads. For example, when performing join operation in join query 3, Spark SQL fetched 23.4 GB data, which is the sum of intermediate data generated from tables scanning, while Shark only fetched 9.6 GB, which is much smaller than the size of data generated by table scanning.

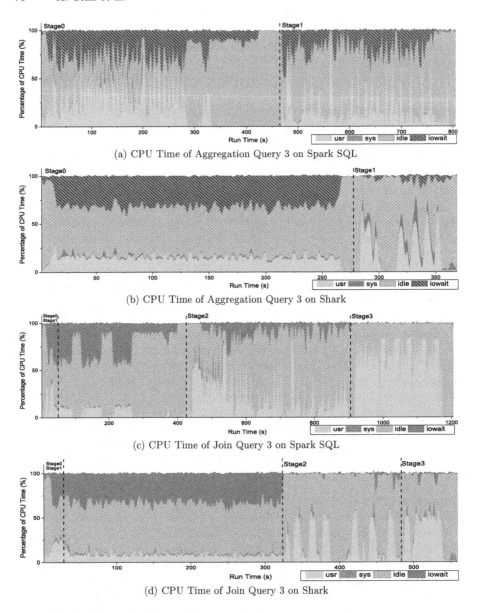

Fig. 3. Percentage of CPU time of different queries on Spark SQL and Shark

We also find that the GC procedure can significantly impact Spark performance. The main reason of the better performance on Shark for aggregation 3 and join 3 is that Shark uses less memory than Spark SQL, which requires less GC operations. However, when the memory is big enough, such as selection queries and aggregation 1, Spark can still perform better than Shark.

Additional Experiment. To verify the correction of our environment results, we also performed another experiment using the database benchmarks from Big-DataBench [14]. Similar to the AMPLab Big Data Benchmark, this benchmark also contains three types of queries, containing select, aggregation and join operations. The difference is that BigDataBench uses two tables of almost same size, which can lead to more shuffle data during complex join and aggregation. These two tables record the information of items and orders respectively. For this additional experiment, we use a 127 GB item table and a 84 GB order table. The result is shown in Fig. 5. The results of select and aggregation have no significant difference, since the aggregation query only performs a simple sum operation, which does not consume much memory during shuffle. The difference appeared in the join query, in which we increased the number of rows filtered from the order table via where condition (50 millions, 800 millions and 1 billion). We can see that Spark SQL performed better than Shark in the first two queries, which is because the memory was still enough. However, the difference decreases as the row number increases, and Spark SQL failed when the row number reached 1 billion, while Shark was able to complete. This result can also indicate that Spark SQL uses more memory than Shark, and can not handle complex join operations on huge data size.

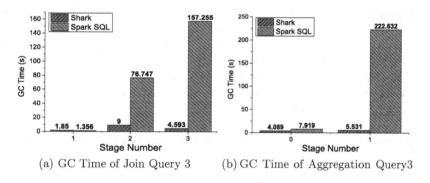

(a) GC Time of Join Query 3   (b) GC Time of Aggregation Query3

Fig. 4. GC time of different queries on Spark SQL and Shark

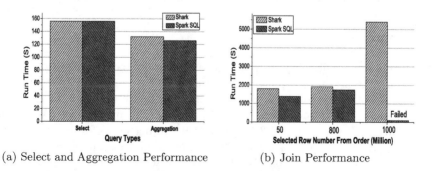

(a) Select and Aggregation Performance   (b) Join Performance

Fig. 5. Query performance with BigDataBench

In summary, the memory usage has become one of main bottlenecks for Spark SQL and more optimizations are needed on such aspect. For users of Spark SQL on current version, we suggest to set memory much bigger and try to make the size of intermediate data smaller via using filter conditions or writing their own optimization rules to Spark SQL. For users whose machines have not enough memory, Shark is still a good choice for complex query processing.

6   Related Works

Researches on the evolution of systems can help both developers and users to get a comprehensive understanding on current systems, and promote the creation of better systems. L. Lu [10] has conducted a comprehensive study of Linux file system evolution. This study focuses on the analysis of patches and code bases of six local file systems, and builds a database for further studies.

Benchmarking big data systems have gained a lot of attention in recent years. Some benchmarks try to evaluate multiple platforms with diversity of workloads. BigDataBench [14] is a big data benchmark suite containing 33 big data benchmarks and 14 real-world data sets, which covers five application domains including search engine, social networks, e-commerce, multimedia analytics and bioinformatics. The workloads of BigDataBench have different implementations include MapReduce, MPI and Spark. There are also some benchmarks that aim at one specific platform. HiBench [6] targets at Hadoop MapReduce and is developed by researchers from Intel. HiBench contains four categories of workloads: Micro Benchmarks, Web Search, Machine Learning, and HDFS Benchmark. SparkBench [9] is a Spark specified benchmark covering four main categories of applications, which includes machine learning, graph computation, SQL query and streaming applications.

There are also works comparing the performance of query processing on different platforms. A. Floratou [5] compared SQL query processing performance of Hive and parallel database system Impala [8] with three SQL workloads, and found Impala was much faster than Hive on MapReduce and Hive on a new engine called Tez [12]. Y. Chen [16] compared five SQL-on-Hadoop systems to provide a comprehensive study on the differences of these system.

7   Conclusion

We present our preliminary work on the evolution of Spark in this paper, which focuses on the study of evolution from Shark to Spark SQL. We demonstrate the changes of Spark cores and SQL parsers from Spark version 0.9.1 to 1.3.1 in details, and also perform a comparison experiment. The experiment results show that Spark SQL with Spark 1.3.1 can not achieve better performance when the size of data raised up to a certain level. The results also indicate Shark has more optimizations than Spark SQL. In future work, we plan to give a more comprehensive study on other types of workloads, and discover more details about the evolution of Spark core components.

Acknowledgements. This work was supported by the National High Technology Research and Development Program of China (Grant No. 2015AA015308), the Major Program of National Natural Science Foundation of China (Grant No. 61432006), and the Key Technology Research and Development Programs of Guangdong Province, China (Grant No. 2015B010108006).

References

1. Big Data Benchmark. https://amplab.cs.berkeley.edu/benchmark/
2. Spark JIRA. https://issues.apache.org/jira/browse/SPARK/
3. Armbrust, M., Xin, R.S., Lian, C., Huai, Y., Liu, D., Bradley, J.K., Meng, X., Kaftan, T., Franklin, M.J., Ghodsi, A., Zaharia, M.: Spark SQL: relational data processing in spark. In: SIGMOD 2015, ACM, New York (2015)
4. Dean, J., Ghemawat, S.: Mapreduce: simplified data processing on large clusters. In: OSDI 2004, USENIX Association, Berkeley (2004)
5. Floratou, A., Minhas, U.F., Özcan, F.: SQL-on-hadoop: Full circle back to shared-nothing database architectures. Proc. VLDB Endowment **7**(12), 1295–1306 (2014)
6. Huang, S., Huang, J., Dai, J., Xie, T., Huang, B.: The hibench benchmark suite: characterization of the mapreduce-based data analysis. In: 2010 IEEE 26th International Conference on Data Engineering Workshops (ICDEW), pp. 41–51. IEEE (2010)
7. Isard, M., Budiu, M., Yu, Y., Birrell, A., Fetterly, D.: Dryad: distributed data-parallel programs from sequential building blocks. In: EuroSys 2007, New York (2007)
8. Kornacker, M., Behm, A., Bittorf, V., Bobrovytsky, T., Ching, C., Choi, A., Erickson, J., Grund, M., Hecht, D., Jacobs, M., et al.: Impala: a modern, open-source SQL engine for hadoop. In: Proceedings of the Conference on Innovative Data Systems Research CIDR 2015 (2015)
9. Li, M., Tan, J., Wang, Y., Zhang, L., Salapura, V.: Sparkbench: a comprehensive benchmarking suite for in memory data analytic platform spark. In: CF 2015, pp. 53:1–53:8. ACM, New York (2015)
10. Lu, L., Arpaci-Dusseau, A.C., Arpaci-Dusseau, R.H., Lu, S.: A study of linux file system evolution. In: FAST 2013, USENIX Association, Berkeley (2013)
11. Melnik, S., Gubarev, A., Long, J.J., Romer, G., Shivakumar, S., Tolton, M., Vassilakis, T.: Dremel: interactive analysis of web-scale datasets. Proc. VLDB Endowment **3**(1–2), 330–339 (2010)
12. Saha, B., Shah, H., Seth, S., Vijayaraghavan, G., Murthy, A., Curino, C.: Apache tez: a unifying framework for modeling and building data processing applications. In: SIGMOD 2015, New York (2015)
13. Thusoo, A., Sarma, J.S., Jain, N., Shao, Z., Chakka, P., Anthony, S., Liu, H., Wyckoff, P., Murthy, R.: Hive: a warehousing solution over a map-reduce framework. Proc. VLDB Endow. **2**(2), 1626–1629 (2009)
14. Wang, L., Zhan, J., Luo, C., Zhu, Y., Yang, Q., He, Y., Gao, W., Jia, Z., Shi, Y., Zhang, S., Zheng, C., Lu, G., Zhan, K., Li, X., Qiu, B.: Bigdatabench: a big data benchmark suite from internet services. In: 20th IEEE International Symposium on High Performance Computer Architecture, HPCA 2014, Orlando, FL, USA, February 15–19, 2014, pp. 488–499 (2014)
15. Xin, R.S., Rosen, J., Zaharia, M., Franklin, M.J., Shenker, S., Stoica, I.: Shark: SQL and rich analytics at scale. In: Proceedings of the 2013 ACM SIGMOD International Conference on Management of Data, pp. 13–24. ACM (2013)

16. Chen, Y., Qin, X., Bian, H., Chen, J., Dong, Z., Du, X., Gao, Y., Liu, D., Lu, J., Zhang, H.: A study of SQL-on-hadoop systems. In: Zhan, J., Rui, H., Weng, C. (eds.) BPOE 2014. LNCS, vol. 8807, pp. 154–166. Springer, Heidelberg (2014)

17. Zaharia, M., Chowdhury, M., Das, T., Dave, A., Ma, J., McCauley, M., Franklin, M.J., Shenker, S., Stoica, I.: Resilient distributed datasets: a fault-tolerant abstraction for in-memory cluster computing. In: NSDI 2012, USENIX Association, Berkeley (2012)

How Data Volume Affects Spark Based Data Analytics on a Scale-up Server

Ahsan Javed Awan[1], Mats Brorsson[1], Vladimir Vlassov[1], and Eduard Ayguade[2](✉)

[1] Software and Computer Systems Department(SCS),
KTH Royal Institute of Technology, Stockholm, Sweden
{ajawan,matsbror,vladv}@kth.se
[2] Computer Architecture Department,
Technical University of Catalunya (UPC), Barcelona, Spain
eduard@ac.upc.edu

Abstract. Sheer increase in volume of data over the last decade has triggered research in cluster computing frameworks that enable web enterprises to extract big insights from big data. While Apache Spark is gaining popularity for exhibiting superior scale-out performance on the commodity machines, the impact of data volume on the performance of Spark based data analytics in scale-up configuration is not well understood. We present a deep-dive analysis of Spark based applications on a large scale-up server machine. Our analysis reveals that Spark based data analytics are DRAM bound and do not benefit by using more than 12 cores for an executor. By enlarging input data size, application performance degrades significantly due to substantial increase in wait time during I/O operations and garbage collection, despite 10 % better instruction retirement rate (due to lower L1 cache misses and higher core utilization). We match memory behaviour with the garbage collector to improve performance of applications between 1.6x to 3x.

Keywords: Scalability · Spark · Micro-architecture

1 Introduction

With a deluge in the volume and variety of data collected, large-scale web enterprises (such as Yahoo, Facebook, and Google) run big data analytics applications using clusters of commodity servers. However, it has been recently reported that using clusters is a case of over-provisioning since a majority of analytics jobs do not process huge data sets and that modern scale-up servers are adequate to run analytics jobs [4]. Additionally, commonly used predictive analytics such as machine learning algorithms work on filtered datasets that easily fit into memory of modern scale-up servers. Moreover the today's scale-up servers can have CPU, memory and persistent storage resources in abundance at affordable prices. Thus we envision small cluster of scale-up servers to be the preferable choice of enterprises in near future.

© Springer International Publishing Switzerland 2016
J. Zhan et al. (Eds.): BPOE 2015, LNCS 9495, pp. 81–92, 2016.
DOI: 10.1007/978-3-319-29006-5_7

While Phoenix [20], Ostrich [6] and Polymer [22] are specifically designed to exploit the potential of a single scale-up server, they don't scale-out to multiple scale-up servers. Apache Spark [21] is getting popular in industry because it enables in-memory processing, scales out to large number of commodity machines and provides a unified framework for batch and stream processing of big data workloads. However it's performance on modern scale-up servers is not fully understood. A recent study [5] characterizes the performance of Spark based data analytics on a scale-up server but it does not quantify the impact of data volume. Knowing the limitations of modern scale-up servers for Spark based data analytics will help in achieving the future goal of improving the performance of Spark based data analytics on small clusters of scale-up servers. In this paper, we answer the following questions concerning Spark based data analytics running on modern scale-up servers:

- Do Spark based data analytics benefit from using larger scale-up servers?
- How severe is the impact of garbage collection on performance of Spark based data analytics?
- Is file I/O detrimental to Spark based data analytics performance?
- How does data size affect the micro-architecture performance of Spark based data analytics?

To answer the above questions, we use empirical evaluation of Apache Spark based benchmark applications on a modern scale-up server. Our contributions are:

- We evaluate the impact of data volume on the performance of Spark based data analytics running on a scale-up server.
- We find the limitations of using Spark on a scale-up server with large volumes of data.
- We quantify the variations in micro-architectural performance of applications across different data volumes.

2   Background

Spark is a cluster computing framework that uses Resilient Distributed Datasets (RDDs) [21] which are immutable collections of objects spread across a cluster. Spark programming model is based on higher-order functions that execute user-defined functions in parallel. These higher-order functions are of two types: Transformations and Actions. Transformations are lazy operators that create new RDDs. Actions launch a computation on RDDs and generate an output. When a user runs an action on an RDD, Spark first builds a DAG of stages from the RDD lineage graph. Next, it splits the DAG into stages that contain pipelined transformations with narrow dependencies. Further, it divides each stage into tasks. A task is a combination of data and computation. Tasks are assigned to executor pool threads. Spark executes all tasks within a stage before moving on to the next stage.

Spark runs as a Java process on a Java Virtual Machine(JVM). The JVM has a heap space which is divided into young and old generations. The young generation keeps short-lived objects while the old generation holds objects with longer lifetimes. The young generation is further divided into eden, survivor1 and survivor2 spaces. When the eden space is full, a minor garbage collection (GC) is run on the eden space and objects that are alive from eden and survivor1 are copied to survivor2. The survivor regions are then swapped. If an object is old enough or survivor2 is full, it is moved to the old space. Finally when the old space is close to full, a full GC operation is invoked.

3   Methodology

3.1   Benchmarks

Table 1 shows the list of benchmarks along with transformations and actions involved. We used Spark versions of the following benchmarks from BigData-Bench [17]. Big Data Generator Suite (BDGS), an open source tool was used to generate synthetic datasets based on raw data sets [15].

- **Word Count (Wc)** counts the number of occurrences of each word in a text file. The input is unstructured Wikipedia Entries.
- **Grep (Gp)** searches for the keyword "The" in a text file and filters out the lines with matching strings to the output file. It works on unstructured Wikipedia Entries.
- **Sort (So)** ranks records by their key. Its input is a set of samples. Each sample is represented as a numerical d-dimensional vector.
- **Naive Bayes (Nb)** uses semi-structured Amazon Movie Reviews data-sets for sentiment classification. We use only the classification part of the benchmark in our experiments.
  K-Means (Km) clusters data points into a predefined number of clusters. We run the benchmark for 4 iterations with 8 desired clusters. Its input is structured records, each represented as a numerical d-dimensional vector.

3.2   System Configuration

Table 2 shows details about our test machine. Hyper-Threading and Turbo-boost are disabled through BIOS because it is difficult to interpret the micro-architectural data with these features enabled [8]. With Hyper-Threading and Turbo-boost disabled, there are 24 cores in the system operating at the frequency of 2.7 GHz.

Table 3 also lists the parameters of JVM and Spark. For our experiments, we use HotSpot JDK version 7u71 configured in server mode (64 bit). The Hotspot JDK provides several parallel/concurrent GCs out of which we use three combinations: (1) Parallel Scavenge (PS) and Parallel Mark Sweep; (2) Parallel New and Concurrent Mark Sweep; and (3) G1 young and G1 mixed for young and old

Table 1. Benchmarks.

Benchmarks		Transformations	Actions
Micro-benchmarks	Word count	map, reduceByKey	saveAsTextFile
	Grep	filter	saveAsTextFile
	Sort	map, sortByKey	saveAsTextFile
Classification	Naive Bayes	map	collect
			saveAsTextFile
Clustering	K-Means	map, filter	takeSample
		mapPartitions	collectAsMap
		reduceByKey	collect

Table 2. Machine details.

Component	Details	
Processor	Intel Xeon E5-2697 V2, Ivy Bridge micro-architecture	
	Cores	12 @ 2.7 GHz (Turbo upto 3.5 GHz)
	Threads	2 per core
	Sockets	2
	L1 Cache	32 KB for instructions and 32 KB for data per core
	L2 Cache	256 KB per core
	L3 Cache (LLC)	30 MB per socket
Memory	2 x 32 GB, 4 DDR3 channels, Max BW 60 GB/s	
OS	Linux kernel version 2.6.32	
JVM	Oracle Hotspot JDK version 7u71	
Spark	Version 1.3.0	

generations respectively. The details on each algorithm are available [2,7]. The heap size is chosen to avoid getting "Out of memory" errors while running the benchmarks. The open file limit in Linux is increased to avoid getting "Too many files open in the system" error. The values of Spark internal parameters after tuning are given in Table 3. Further details on the parameters are available [3].

3.3   Measurement Tools and Techniques

We configure Spark to collect GC logs which are then parsed to measure time (called real time in GC logs) spent in garbage collection. We rely on the log files generated by Spark to calculate the execution time of the benchmarks. We use Intel Vtune [1] to perform concurrency analysis and general micro-architecture exploration. For scalability study, each benchmark is run 5 times within a single JVM invocation and the mean values are reported. For concurrency analysis, each benchmark is run 3 times within a single JVM invocation and Vtune measurements are recorded for the last iteration. This experiment is repeated 3 times

Table 3. JVM and Spark parameters for different workloads.

		Wc	Gp	So	Km	Nb
JVM	Heap Size (GB)	50				
	Old Generation Garbage Collector	PS MarkSweep				
	Young Generation Garbage Collector	PS Scavange				
Spark	spark.storage.memoryFraction	0.1	0.1	0.1	0.6	0.1
	spark.shuffle.memoryFraction	0.7	0.7	0.7	0.4	0.7
	spark.shuffle.consolidateFiles	true				
	spark.shuffle.compress	true				
	spark.shuffle.spill	true				
	spark.shuffle.spill.compress	true				
	spark.rdd.compress	true				
	spark.broadcast.compress	true				

and the best case in terms of execution time of the application is chosen. The same measurement technique is also applied in general architectural exploration, however the difference is that mean values are reported. Additionally, executor pool threads are bound to the cores before collecting hardware performance counter values.

We use the top-down analysis method proposed by Yasin [18] to study the micro-architectural performance of the workloads. Super-scalar processors can be conceptually divided into the "front-end" where instructions are fetched and decoded into constituent operations, and the "back-end" where the required computation is performed. A pipeline slot represents the hardware resources needed to process one micro-operation. The top-down method assumes that for each CPU core, there are four pipeline slots available per clock cycle. At issue point each pipeline slot is classified into one of four base categories: Front-end Bound, Back-end Bound, Bad Speculation and Retiring. If a micro-operation is issued in a given cycle, it would eventually either get retired or cancelled. Thus it can be attributed to either Retiring or Bad Speculation respectively. Pipeline slots that could not be filled with micro-operations due to problems in the front-end are attributed to Front-end Bound category whereas pipeline slot where no micro-operations are delivered due to a lack of required resources for accepting more micro-operations in the back-end of the pipeline are identified as Back-end Bound.

4   Scalability Analysis

4.1   Do Spark Based Data Analytics Benefit from Using Scale-up Servers?

We configure spark to run in local-mode and used system configuration parameters of Table 3. Each benchmark is run with 1, 6, 12, 18 and 24 executor pool threads. The size of input data-set is 6 GB. For each run, we set the CPU affinity of the Spark process to emulate hardware with same number of cores as the worker threads. The cores are allocated from one socket first before switching to the second socket. Figure 1a plots speed-up as a function of the number of cores. It shows that benchmarks scale linearly up to 4 cores within a socket. Beyond 4 cores, the workloads exhibit sub-linear speed-up, e.g., at 12 cores within a socket, average speed-up across workloads is 7.45. This average speed-up increases up to 8.74, when the Spark process is configured to use all 24 cores in the system. The performance gain of mere 17.3 % over the 12 cores case suggest that Spark applications do not benefit significantly by using more than 12-core executors.

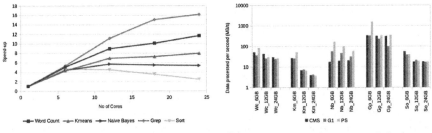

(a) Benchmarks do not benefit by adding more than 12 cores.

(b) Data processed per second decreases with increase in data size.

Fig. 1. Scale-up performance of applications: (a) when the number of cores increases and (b) when input data size increases.

4.2 Does Performance Remain Consistent as We Enlarge the Data Size?

The benchmarks are configured to use 24 executor pool threads in the experiment. Each workload is run with 6 GB, 12 GB and 24 GB of input data and the amount of data processed per second (DPS) is calculated by dividing the input data size by the total execution time. The data sizes are chosen to stress the whole system and evaluate the system's data processing capability. In this regard, DPS is a relevant metric as suggested in by Luo et al. [14]. We also evaluate the sensitivity of DPS to garbage collection schemes but explain it in the next section. Here we only analyse the numbers for Parallel Scavenge garbage collection scheme. By comparing 6 GB and 24 GB cases in Fig. 1b, we see that K-Means performs the worst as its DPS decreases by 92.94 % and Grep performs the best with a DPS decrease of 11.66 %. Furthermore, we observe that DPS decreases by 49.12 % on average across the workloads, when the data size is increased from 6 GB to 12 GB. However DPS decreases further by only 8.51 % as the data size is increased to 24 GB. In the next section, we will explain the reason for poor data scaling behaviour.

5 Limitations to Scale-up

5.1 How Severe is the Impact of Garbage Collection?

Because of the in-memory nature of most Spark computations, garbage collection can become a bottleneck for Spark programs. To test this hypothesis, we analysed garbage collection time of scalability experiments from the previous section. Figure 2a plots total execution time and GC time across the number of cores. The proportion of GC time in the execution time increases with the number of cores. At 24 cores, it can be as high as 48 % in K-Means. Word Count and Naive Bayes also show a similar trend. This shows that if the GC time had at least not been increasing, the applications would have scaled better. Therefore we conclude that GC acts as a bottleneck.

To answer the question, "How does GC affect data processing capability of the system?", we examine the GC time of benchmarks running at 24 cores. The input data size is increased from 6 GB to 12 GB and then to 24 GB. By comparing 6 GB and 24 GB cases in Fig. 2b, we see that GC time does not increase linearly, e.g., when input data is increased by 4x, GC time in K-Means increases by 39.8x. A similar trend is also seen for Word Count and Naive Bayes. This also shows that if GC time had been increasing at most linearly, DPS would not have decreased significantly. For K-Means, DPS decreases by 14x when data size increases by 4x. For similar scenario in Naive Bayes, DPS decreases by 3x and GC time increases by 3x. Hence we can conclude that performance of Spark applications degrades significantly because GC time does not scale linearly with data size.

Finally we answer the question, "Does the choice of Garbage Collector impact the data processing capability of the system?". We look at impact of three garbage collectors on DPS of benchmarks at 6 GB, 12 GB and 24 GB of input data size. We study out-of-box (without tuning) behaviour of Concurrent Mark Sweep, G1 and Parallel Scavenge garbage collectors. Figure 2b shows that across all the applications, GC time of Concurrent Mark Sweep is the highest and GC time of Parallel Scavenge is the lowest among the three choices. By comparing

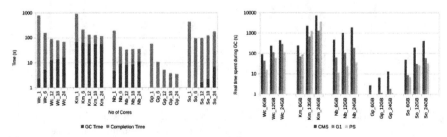

(a) GC overhead is a scalability bottleneck. (b) GC time increases at a higher rate with data size.

Fig. 2. Impact of garbage collection on application performance: (a) when the number of cores increases and (b) when input data size increases.

the DPS of benchmarks across different garbage collectors, we see that Parallel Scavenge results in 3.69x better performance than Concurrent Mark Sweep and 2.65x better than G1 on average across the workloads at 6 GB. At 24 GB, Parallel Scavenge performs 1.36x better compared to Concurrent Mark Sweep and 1.69x better compared to G1 on average across the workloads.

5.2 Does File I/O Become a Bottleneck Under Large Data Volumes?

In order to find the reasons for poor performance of Spark applications under larger data volumes, we studied the thread-level view of benchmarks by performing concurrency analysis in Intel Vtune. We analyse only executor pool threads as they contribute to 95 % of total CPU time during the entire run of the workloads. Figure 3b shows that CPU time and wait time of all executor pool threads. CPU time is the time during which the CPU is actively executing the application on all cores. Wait time occurs when software threads are waiting on I/O operations or due to synchronization. The wait time is further divided into idle time and wait on file I/O operations. Both idle time and file I/O time are approximated from the top 5 waiting functions of executor pool threads. The remaining wait time comes under the category of "other wait time".

It can be seen that the fraction of wait time increases with increase in input data size, except in Grep where it decreases. By comparing 6 GB and 24 GB case, the data shows that the fraction of CPU time decreases by 54.15 %, 74.98 % and 82.45 % in Word Count, Naive Bayes and Sort respectively; however it increases by 21.73 % in Grep. The breakdown of wait time reveals that contribution of file I/O increases by 5.8x, 17.5x and 25.4x for Word Count, Naive Bayes and Sort respectively but for Grep, it increases only 1.2x. The CPU time in Fig. 3b also correlates with CPU utilization numbers in Fig. 3a. On average across the workloads, CPU utilization decreases from 72.34 % to 39.59 % as the data size is increased from 6 GB to 12 GB which decreases further by 5 % in 24 GB case.

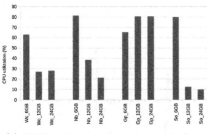

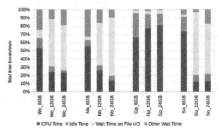

(a) CPU utilization decreases with data size.

(b) Wait time becomes dominant at larger datasets due to significant increase in file I/O operations.

Fig. 3. Time breakdown under executor pool threads.

5.3 Is Micro-architecture Performance Invariant to Input Data Size?

We study the top-down breakdown of pipeline slots in the micro-architecture using the general exploration analysis in Vtune. The benchmarks are configured to use 24 executor pool threads. Each workload is run with 6 GB, 12 GB and 24 GB of input data. Figure 4a shows that benchmarks are back-end bound. On average across the workloads, retiring category accounts for 28.9 % of pipeline slots in 6 GB case and it increases to 31.64 % in the 24 GB case. Back-end bound fraction decreases from 54.2 % to 50.4 % on average across the workloads. K-Means sees the highest increase of 10 % in retiring fraction in 24 GB case in comparison to 6 GB case.

Next, we show the breakdown of memory bound stalls in Fig. 4b. The term DRAM Bound refers to how often the CPU was stalled waiting for data from main memory. L1 Bound shows how often the CPU was stalled without missing in the L1 data cache. L3 Bound shows how often the CPU was stalled waiting for the L3 cache, or contended with a sibling core. Store Bound shows how often the CPU was stalled on store operations. We see that DRAM bound stalls are the primary bottleneck which account for 55.7 % of memory bound stalls on average across the workloads in the 6 GB case. This fraction however decreases to 49.7 % in the 24 GB case. In contrast, the L1 bound fraction increase from 22.5 % in 6 GB case to 30.71 % in 24 GB case on average across the workloads.

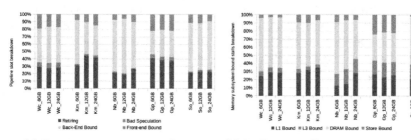

(a) Retiring rate increases at larger datasets.

(b) L1 Bound stalls increase with data size.

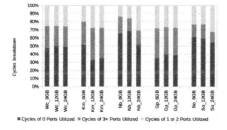

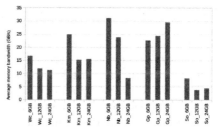

(c) Port utilization increases at larger datasets.

(d) Memory traffic decreases with data size.

Fig. 4. Micro-architecture performance is inconsistent across different data sizes.

It means that due to better utilization of L1 cache, the number of simultaneous data read requests to the main memory controller decreases at larger volume of data. Figure 4d shows that average memory bandwidth consumption decreases from 20.7 GB/s in the 6 GB case to 13.7 GB/s in the 24 GB case on average across the workloads.

Figure 4c shows the fraction of cycles during execution ports are used. Ports provide the interface between instruction issue stage and the various functional units. By comparing 6 GB and 24 GB cases, we observe that cycles during which no port is used decrease from 51.9 % to 45.8 % on average across the benchmarks and cycles during which 1 or 2 ports are utilized increase from 22.2 % to 28.7 % on average across the workloads.

6   Related Work

Several studies characterize the behaviour of big data workloads and identify the mismatch between the processor and the big data applications [9–13,17,19]. However these studies lack in identifying the limitations of modern scale-up servers for Spark based data analytics. Ousterhout et al. [16] have developed blocked time analysis to quantify performance bottlenecks in the Spark framework and have found out that CPU and not I/O operations are often the bottleneck. Our thread level analysis of executor pool threads shows that the conclusion made by Ousterhout et al. is only valid when the input data-set fits in each node's memory in a scale-out setup. When the size of data set on each node is scaled-up, file I/O becomes the bottleneck again. Wang et al. [17] have shown that the volume of input data has considerable affect on the micro-architecture behaviour of Hadoop based workloads. We make similar observation about Spark based data analysis workloads.

7   Conclusions

We have reported a deep dive analysis of Spark based data analytics on a large scale-up server. The key insights we have found are as follows:

- Spark workloads do not benefit significantly from executors with more than 12 cores.
- The performance of Spark workloads degrades with large volumes of data due to substantial increase in garbage collection and file I/O time.
- With out any tuning, Parallel Scavenge garbage collection scheme outperforms Concurrent Mark Sweep and G1 garbage collectors for Spark workloads.
- Spark workloads exhibit improved instruction retirement due to lower L1 cache misses and better utilization of functional units inside cores at large volumes of data.
- Memory bandwidth utilization of Spark benchmarks decreases with large volumes of data and is 3x lower than the available off-chip bandwidth on our test machine.

We conclude that Spark run-time needs node-level optimizations to maximize its potential on modern servers. Garbage collection is detrimental to performance of in-memory big data systems and its impact could be reduced by careful matching of garbage collection scheme to workload. Inconsistencies in microarchitecture performance across the data sizes pose additional challenges for computer architects. Off-chip memory buses should be optimized for in-memory data analytics workloads by scaling back unnecessary bandwidth.

References

1. Intel Vtune Amplifier XE 2013. http://software.intel.com/en-us/node/544393
2. Memory manangement in the java hotspot virtual machine. http://www.oracle.com/technetwork/java/javase/memorymanagement-whitepaper-150215.pdf
3. Spark configuration. https://spark.apache.org/docs/1.3.0/configuration.html
4. Appuswamy, R., Gkantsidis, C., Narayanan, D., Hodson, O., Rowstron, A.I.T.: Scale-up vs scale-out for hadoop: time to rethink? In: ACM Symposium on Cloud Computing, SOCC, p. 20 (2013)
5. Awan, A.J., Brorsson, M., Vlassov, V., Ayguadé, E.: Performance characterization of in-memory data analytics on a modern cloud server. arXiv preprint arXiv:1506.07742 (2015)
6. Chen, R., Chen, H., Zang, B.: Tiled-mapreduce: Optimizing resource usages of data-parallel applications on multicore with tiling. In: Proceedings of the 19th International Conference on Parallel Architectures and Compilation Techniques, pp. 523–534. PACT 2010 (2010)
7. Detlefs, D., Flood, C., Heller, S., Printezis, T.: Garbage-first garbage collection. In: Proceedings of the 4th international symposium on Memory management, pp. 37–48. ACM (2004)
8. Levinthal, D.: Performance analysis guide for intel core i7 processor and intel xeon 5500 processors. In: Intel Performance Analysis Guide (2009)
9. Ferdman, M., Adileh, A., Kocberber, O., Volos, S., Alisafaee, M., Jevdjic, D., Kaynak, C., Popescu, A.D., Ailamaki, A., Falsafi, B.: Clearing the clouds: A study of emerging scale-out workloads on modern hardware. In: Proceedings of the Seventeenth International Conference on Architectural Support for Programming Languages and Operating Systems, pp. 37–48. ASPLOS XVII (2012)
10. Jia, Z., Wang, L., Zhan, J., Zhang, L., Luo, C.: Characterizing data analysis workloads in data centers. In: IEEE International Symposium on Workload Characterization (IISWC), pp. 66–76 (2013)
11. Jia, Z., Zhan, J., Wang, L., Han, R., McKee, S.A., Yang, Q., Luo, C., Li, J.: Characterizing and subsetting big data workloads. In: IEEE International Symposium on Workload Characterization (IISWC), pp. 191–201 (2014)
12. Jiang, T., Zhang, Q., Hou, R., Chai, L., McKee, S.A., Jia, Z., Sun, N.: Understanding the behavior of in-memory computing workloads. In: IEEE International Symposium on Workload Characterization (IISWC), pp. 22–30 (2014)
13. Karakostas, V., Unsal, O.S., Nemirovsky, M., Cristal, A., Swift, M.: Performance analysis of the memory management unit under scale-out workloads. In: IEEE International Symposium on Workload Characterization (IISWC), pp. 1–12, October 2014

14. Luo, C., Zhan, J., Jia, Z., Wang, L., Lu, G., Zhang, L., Xu, C.Z., Sun, N.: Cloudrank-d: Benchmarking and ranking cloud computing systems for data processing applications. Front. Comput. Sci. **6**(4), 347–362 (2012)

15. Ming, Z., Luo, C., Gao, W., Han, R., Yang, Q., Wang, L., Zhan, J.: BDGS: A scalable big data generator suite in big data benchmarking. In: Rabl, T., Raghunath, N., Poess, M., Bhandarkar, M., Jacobsen, H.-A., Baru, C. (eds.) Advancing Big Data Benchmarks. LNCS, pp. 138–154. Springer, Heidelberg (2014)

16. Ousterhout, K., Rasti, R., Ratnasamy, S., Shenker, S., Chun, B.G.: Making sense of performance in data analytics frameworks. In: 12th USENIX Symposium on Networked Systems Design and Implementation (NSDI 2015), pp. 293–307 (2015)

17. Wang, L., Zhan, J., Luo, C., Zhu, Y., Yang, Q., He, Y., Gao, W., Jia, Z., Shi, Y., Zhang, S., Zheng, C., Lu, G., Zhan, K., Li, X., Qiu, B.: Bigdatabench: A big data benchmark suite from internet services. In: 20th IEEE International Symposium on High Performance Computer Architecture, HPCA, pp. 488–499 (2014)

18. Yasin, A.: A top-down method for performance analysis and counters architecture. In: 2014 IEEE International Symposium on Performance Analysis of Systems and Software, ISPASS, pp. 35–44 (2014)

19. Yasin, A., Ben-Asher, Y., Mendelson, A.: Deep-dive analysis of the data analytics workload in cloudsuite. In: IEEE International Symposium on Workload Characterization (IISWC), pp. 202–211, October 2014

20. Yoo, R.M., Romano, A., Kozyrakis, C.: Phoenix rebirth: Scalable mapreduce on a large-scale shared-memory system. In: Proceedings of IEEE International Symposium on Workload Characterization (IISWC), pp. 198–207 (2009)

21. Zaharia, M., Chowdhury, M., Das, T., Dave, A., Ma, J., McCauly, M., Franklin, M.J., Shenker, S., Stoica, I.: Resilient distributed datasets: A fault-tolerant abstraction for in-memory cluster computing. In: Presented as part of the 9th USENIX Symposium on Networked Systems Design and Implementation (NSDI 2012), pp. 15–28. San Jose, CA (2012)

22. Zhang, K., Chen, R., Chen, H.: Numa-aware graph-structured analytics. In: Proceedings of the 20th ACM SIGPLAN Symposium on Principles and Practice of Parallel Programming, pp. 183–193. ACM (2015)

An Optimal Reduce Placement Algorithm for Data Skew Based on Sampling

Zhuo Tang[1]([⊠]), Wen Ma[1], Rui Li[1], Kenli Li[1], and Keqin Li[1,2]

[1] College of Computer Science and Electronic Engineering,
Hunan University, Changsha 410082, China
{ztang,lkl}@hnu.edu.cn, mawen1029@126.com
[2] The Department of Computer Science, State University of New York,
New Paltz, NY 12561, USA
lik@newpaltz.edu

Abstract. For frequent disk I/O and big data transmissions among different racks and physical nodes, the intermediate data communication has become the biggest performance bottle-neck in most running Hadoop systems. This paper proposes a reduce placement algorithm called CORP to schedule related map and reduce tasks on the near nodes or clusters or racks for the data locality. Since the number of keys cannot be counted until the input data are processed by map tasks, this paper firstly provides a sampling algorithm based on reservoir sampling to achieve the distribution of the keys in intermediate data. Through calculating the distance and cost matrices among the cross node communication, the related map and reduce tasks can be scheduled to relatively near physical nodes for data locality. Experimental results show that CORP can not only improve the balance of reduce tasks effectively, but also decrease the job execution time for the lower inner data communication.

Keywords: Data sampling · Data skew · Inner communication · MapReduce · Reduce placement

1 Introduction

In Hadoop framework, because map tasks always output the intermediate data in the local nodes, the data should be transmitted from the map nodes to corresponding reduce nodes, which is an all-to-all communication model. The frequent disk I/O and big data transmissions have become the biggest performance bottle-neck in most running Hadoop systems, which may saturate the top-of-rack switch and inflate job execution time [1]. Cross-rack communication happens if a mapper and a reducer reside in different racks, which is very often in todays data center environments [2]. Due to the limitation of the switches in clusters, the overall performance of a system is often not satisfactory when working on large data sets [3].

To mitigate network traffic, it is an effective method to place reduce tasks close to the physical nodes on which map tasks generate intermediate data used

© Springer International Publishing Switzerland 2016
J. Zhan et al. (Eds.): BPOE 2015, LNCS 9495, pp. 93–106, 2016.
DOI: 10.1007/978-3-319-29006-5_8

as reduce input [4]. Because the intermediate keys distribution is the determining factor for the input data distribution of reduce tasks, if the intermediate data from map tasks are distributed to reduce tasks uniformly, reduce locality and task placement are not able to optimize the all-to-all communication in Hadoop. But luckily, data skew is universal existence in all input data. Current researches proving that moving task is more efficient than moving data in the Hadoop distributed environment [5,6], where data skews are widespread (some key values are significantly more frequent than others). These studies make it possible for us to solve the cross-racks/nodes communication problem through reduce task placement. To tackle this problem, this paper provides a dynamic range partition method which conducts a pre-run sample of the input before the real job.

Existing researches for reduce tasks requesting intermediate outputs are not only without any data locality consideration, but also partitioning skew unaware. Most versions of Hadoop usually employ static hash functions to partition the intermediate data. This works well when the data are uniformly distributed, but can perform poorly when the intermediate results of input are skewed. In the presence of data skew, this paper proposes a communication oriented reduce placement (CORP) method to decrease all-to-all communications between mappers and reducers, whose basic idea is to place related map and reduce tasks on the near nodes or clusters or racks. Since data skew is difficult to solve if the input distribution is unknown, a normal thought is to examine the data before deciding the partition. In a real application, the intermediate outputs can be monitored and counted only after a job begins running, but it is meaningless to obtain the key value distribution after processing the whole input data.

To tackle this problem, this paper provides a dynamic range partition method which conducts a pre-run sample of the input before the real job. By integrating sampling into a small percentage of the map tasks, this paper prioritizes the execution of sampling tasks over the normal map tasks in order to achieve the distribution statistics. The main contributions of this paper are summarized below:

- We propose a new sampling method for large Hadoop input data. With the random data replacement policy in the sampling zone, this sampling algorithm can achieve a much better approximation to the distribution of the intermediate data.
- We propose a novel reduce placement algorithm based on data distribution, which can schedule the related map and reduce tasks on the near nodes for the data locality. This algorithm can reduce the all-to-all communication among inner Hadoop cluster.
- We implement CORP in Hadoop 2.4.0 and evaluate its performance for some most common benchmarks. Experiment results show that CORP can decrease the data transmission on core switch by 21.45 % comparing to default hash mechanism.

The rest of the paper is organized as follows: Sect. 2 surveys related works on reducer placement and data skew. Section 3 proposes data sampling algorithm of MapReduce framework. Section 4 proposes the reduce placement algorithm. Performance evaluation is in Sect. 5. Section 6 concludes the paper.

2   Background and Related Works

There are two optimal algorithms to solve the reducer placement problem (RPP), and an analytical method to find the minimum (may not be feasible) solution of RPP, which considers the placement of reducers to minimize cross-rack traffic. One algorithm is a greedy algorithm [7], which assigns one reduce task to a rack at a time. When assigning a reduce task to a rack, it chooses the rack which incurs the minimum total traffic (up and down) if the reduce task is assigned to that rack. The second algorithm, called binary search [8], uses binary search to find the minimum bound of the traffic function for each rack, and then uses that minimum bound to find the number of reducers on each rack. They used the idle network resource to fetch the next key-value pair when working on current key-value pair.

Actually, data skew is not a new problem specific to MapReduce. For the distribution of intermediate data in Hadoop is typically skewed, we need to face many real world applications exhibiting large amount of data skew, including scientific applications [9], distributed database operations like join, grouping and aggregation [10], search engine applications (Page Rank, Inverted Index, etc.) and some simple applications (sort, grep, etc.) [11]. Ardizzoni [12] has witnessed the data skew phenomenon in the Microsoft production cluster. How to handle data-skew effects has been studied previously in the parallel database researches [13], but still lacks an effective prediction model for the distribution of the intermediate keys.

There are three typical skews in MapReduce: skewed key frequencies, skewed tuple sizes, and skewed execution times [14]. In this paper, we mainly focus on the first type. If some keys appear more frequently in the intermediate data tuples, the number of tuples per cluster owned will be different. Even if every reducer receives the same number of clusters, the overall number of tuples per reducer received will be different. The skewed key frequencies is a necessary condition for using the reduce placement polices to minimize cross-rack traffic, and it is meaningless to place the reduce tasks if the key frequencies of intermediate data are evenly distributed.

Chen et al. [15] presented LIBRA, a lightweight strategy to solve the data skew problem for reduce-side applications in MapReduce. LIBRA estimates the intermediate data distribution through sampling the partial map tasks, and uses an innovative approach to balance the load among the reduce tasks, which supports the split of large keys. Comparing to these previous works, our goal of the reduce placement algorithm proposed in this paper is to reduce the all-to-all communication among inner Hadoop cluster based on a more accurate estimation for the intermediate data skew.

Hammoud et al. [18] and Chen et al. [17] presented a locality-aware scheduling of MapReduce. They are focused on diminishing network traffic and improve performance in data parallel systems,but only collocating reduce tasks with the maximum data that may not pull the suitable intermediate results and being a poor prediction mechanism after recognizing input data network locations and sizes.

3 Data Sampling in MapReduce Framework

3.1 Data Skew Model

In this model, for quantifying the data received by a special reduce task, some initial and intermediate results with their relationships can be formalized as follows.

1. C: A matrix of $n \times p$ which defines the distribution of intermediate results in each partition. C_{ij} denotes the number key/value tuples processed by j^{th} reducer from i^{th} node, and n-1 is the number of various nodes. If $C_{ij} = k$, the number of the key/value tuples from the node n_i is k in the partition P_j, which will be allocated to the reduce task R_k. In this paper, the number of the partitions is treated equal the reduce amount here.

2. $C^{\sigma,l}$: A matrix of $m \times p$ which defines the number of map tasks where the key/value tuples will be allocated to reduce tasks by the node l. Under normal conditions, the key number of original input data follows Zipf distributions [16]. To select a good ratio, we use a varying parameters (from 0.1 to 1.2) to control the degree of the skew. Larger value means heavier skew, and it also determines the distribution of C_{ij}. In this model, the number of key/value pairs that the reducer j contains is denoted as $RC(j)$. Without loss of generality, the value of $RC(j)$ with a skew degree could be defined as follows:

$$RC(j, \sigma) = \sum_{i=0}^{n-1} C_{i,j}^{\sigma} = \sum_{i=0}^{n-1} \sum_{k=0}^{m-1} C_{k,j}^{\sigma,i} \tag{1}$$

And on this basis, we can calculate the average number of key/value tuples of all running reduce tasks as follows, in which the parameter m is the number of reduce tasks:

$$\overline{mean_{\sigma}} = \frac{\sum_{i=0}^{n-1} RC(j, \sigma)}{p} = \frac{\sum_{j=0}^{p-1} \sum_{i=0}^{n-1} \sum_{k=0}^{m-1} C_{k,j}^{\sigma,i}}{p} \tag{2}$$

The intermediate data processed by a reduce task are considered skew when the following condition is satisfied:

$$|\overline{mean_{\sigma}} - RC(j, \sigma)| > std \tag{3}$$

In Eq. (3), std is a standard deviation of the number of key/value tuples for all reduce tasks, which can be used to measure the overall load balancing level of reducers. The value of this standard deviation for all intermediate results in reduce tasks can be calculated by Eq. (4).

$$std = \sqrt{\frac{\sum_{k=0}^{p-1} \left(\frac{\sum_{j=0}^{p-1} \sum_{i=0}^{n-1} C_{i,j}^{\sigma}}{n} - \sum_{i=0}^{n-1} C_{i,k}^{\sigma} \right)^2}{p}} \cdot \tag{4}$$

Hence, we can calculate the difference between the average intermediate results of all reduce tasks and the number of key/value pairs belong to k^{th} reducer as follows:

$$\overline{mean_\sigma} - RC(v,\sigma) = \frac{\sum_{j=0}^{p-1}\sum_{i=0}^{n-1} C_{i,j}^\sigma}{p} - \sum_{i=0}^{n-1} C_{i,v}^\sigma = \frac{\sum_{j=0}^{p-1}\sum_{i=0}^{n-1}\sum_{k=0}^{m-1} C_{k,j}^{\sigma,i} - p\sum_{i=0}^{n-1}\sum_{k=0}^{m-1} C_{k,v}^{\sigma,i}}{p}$$

$$(5)$$

In this case, when a reduce task is load balancing, $|mean - RC(k)| < std$ is always satisfied. As a result, when the number of key/value tuples assigned to reducer k is larger than the value of mean, the kth reducer will be taken as a skew task even though it is running normally.

To measure the data skew degree of all the reduce tasks, this paper uses an indicator Fos (Factor of Skew) to quantize data skew and load balancing:

$$Fos = std(C^{\sigma,l})/\overline{mean_\sigma} \qquad (6)$$

The smaller value of Fos, the better the load balancing, and lower data skew will be obtained.

3.2   Data Sampling Algorithm

The Sampler Implementation in the Hadoop Systems. To ascertain the distribution of the intermediate data is the only way to make a reduce placement strategy and calculating the optimal solution to the above problem is unrealistic.The sampling must start up before the input data processed.

Therefore, we present a distributed approximation algorithm by sampling and estimation. Our sample strategies are also running through invoking this InputSampler class in underlying implementation.

Reservoir Sampling Algorithm. From $(k + 1)^{th}$ chunk, the ith chunk will be selected with the probability: $1/i$ $(i = k + 1, k + 2, \cdots, N)$, and the chunk will appear in the *reservoir* with probability: k/i. In this process, an element in the original *reservoir* will be replaced randomly, until all of the k chunks in *reservoir* are completely substituted. After one traversal, we can get entirely random k chunks from the original input data. It can be proved that each data will be taken out with equal probability k/n when the unknowable amount of whole input data equals to n, $n \geq k$.

From Algorithm 1, because the intermediate tuples distribution of sample data keeps coherent with the whole input data, we can calculate the data size of reduce tasks from every map node under the consistent distribution law. Algorithm 2 provides the process to obtain the distribution of the intermediate tuples for each reduce task. For a specific MapReduce job, this algorithm firstly starts this job with the sample data, and records the number of tuples from a map local node to each reduce task based on a monitor in each map node.

Algorithm 1. Reservoir Sampling

Require:
  The *reservoir* size: r.
Ensure:
  The sample data block set BS.
  set BS as a set of the sample data block, $BS=\varnothing$;
  $k = 0$;
  for each block b in input data **do**
    $k=k+1$;
    if $k \leq r$ **then**
      $BS = BS \bigcup b$;
    else
      select a block b' in input data;
      replace a randomly selected block in the *reservoir* with the block b';
    end if
  end for
  return BS.

Function *getOrignalMapNode* is used to retrace the intermediate tuples, and obtain the map nodes which produce these data [19]. In this section,the sampling policy through marking the location of splits instead of moving the practical data. Obviously, this is a trade-off between the sampling overhead and the accuracy of the result.

4 Communication Oriented Reduce Placement

4.1 The Model of MapReduce

In this section, we describe the principles of our design and provide a formal analysis of the problem. Let n be the number of nodes and m be the number of map tasks. We firstly initialize the node set as $\{N_0, N_1, \cdots, N_{n-1}\}$ and map set as $\{M_0, M_1 \cdots, M_{m-1}\}, 0 \leq n \leq m$ for rack set $R_r, r \in \{0, 1, \cdots, k\}, 0 \leq k \leq n-1$. For this model, some specific data structures can be formalized as follows:

(1) V: A vector of length p whose elements indicate the relevant number of key/value tuples in every node. If $v_{ij} = k$, there are k key/value pairs in node N_i assigned to the reduce j. Therefore:

$$v_{lj} = \sum_{i=0}^{m} C_{i,j}^l, 0 < l < n \qquad V_l = [v_{l0}, v_{l1}, \cdots, v_{lp}], C = [V_0, V_1, \cdots, V_n] \qquad (7)$$

(2) D: A matrix of $n \times n$ which defines the distance between physical nodes. According the network latency, we can define the distance between two physical nodes in the same rack as d_1, similarly, in different racks but in the same cluster as d_2, in different clusters as d_3, in different datacenters as d_4,in the same node as 0.The distance value would increase with the rising physical distances: $0 < d_1 < d_2 < d_3 < d_4$. This paper supposes that the shorter the distance,

---

Algorithm 2. Distribution Detection

Require:
    The sample of data blocks BS;
    a MapReduce job: mrj;
    the number of computing nodes N;
    the number of reduce tasks R.
Ensure:
    The matrix C of whole input data.
  1: run the MapReduce job mrj using BS as the input data;
  2: initialize an intermediate matrix C'
      $C'_{ij} = 0,\ 0 < i \leq N,\ 0 < j \leq R$;
  3: **for** each cluster c_k with tuple key k **do**
  4:    calculate the sequence number j of reduce which c_k should be send to;
  5:    $j = systemhash(k)$;
  6:    **for** each map node i **do**
  7:      **if** $i = getOrignalMapNode(c_k)$ **then**
  8:        $C'_{ij} = C'_{ij} + tuple\_number\_of(c_k)$;
  9:      **end if**
10:    **end for**
11: **end for**
12: SN = the number of tuples in BS;
13: WN = the number of tuples in whole input data;
14: **for** each map node i **do**
15:    **for** each reduce task j do **do**
16:      $C_{ij} = (C'_{ij} \times WN) / SN$;
17:    **end for**
18: **end for**
19: **return** matrix $C = \{C_{ij}\}$.

---

the faster the data transferring speed, which is the theoretical basis of the model optimization.

(3) R: A matrix of $r \times n$ which defines the position of reduce tasks started on the node, and is a typical sparse matrix. The element r_{ji} is a boolean variable which indicates whether the reduce task i is set up on the node j. The following condition ensures that each reduce task should be placed on only one node:

$$\sum_i r_{ji} = 1, \forall j \in [0, n-1] \tag{8}$$

On this basis, we use a vector RV_j to denote the position of a task on a node in the matrix R. The element r_{ji} in vector RV_j is a boolean variable which indicates whether the reduce task j is placed on the node i,Therefore, if there are r reduce tasks among nodes, we can define the matrix R as follows:

$$RV_j = [r_{j0}, r_{j1}, \cdots, r_{jn}];\ R = [RV_0, RV_1, \cdots, RV_r]^T \tag{9}$$

To capture locality, we define a communication cost function that measures data transferring between two nodes.Using the network hops as the near-far measure, this paper defines a vector DV to calculate the distance among node N_i and other nodes:

$$DVi = [dis_1, dis_2, \cdots, dis_n], 0 < |dis_i| < d_4 \tag{10}$$

For the process of reduce tasks fetching the corresponding data, there are two key factors about cost matrix T: the amount of data transferred and the distances between the map nodes and reduce nodes. An element T_{ij} of the cost matrix can be obtained as the follows:

$$t_{ij} = DV_i \times TV_j^T = \sum_{d,i'=1}^{n} dis_{id} \times r_{i'd} \qquad \forall i, j \in [0, n-1] \tag{11}$$

In the above discussion, C^l is defined as an intermediate results allocation matrix, which represents the key/value pairs distribution in node l. And $t_{i,j}^k$ denotes k^{th} placement choice for reduce j on node k. The minimal cost MC of the whole Hadoop system can be calculated through multiplying the distribution matrix C by the cost matrix T from a distinguished central node:

$$MC\left(C^{\sigma,l}, T\right) = \min_k \left(\sum_{i=0}^{m-1} \left[\sum_{j=0}^{p-1} c_{ij}^{\sigma,l} \right] \times t_{ij}^k \right) \tag{12}$$

From Eq. (12), in the data skew environment, if most data for a reduce task j come from the node i, when MC gets the minimal value, the jth reduce task can fetch the largest data local blocks in k^{th} computing node. To achieve an optimal reduce tasks placement solution which can minimize the network communication overhead among different physical nodes, the objective function of this model can be specified as the following optimization problem:

$$Minimize \sum_{m,n} MC(C^l, T) \tag{13}$$

Hence, the reduce task placement can be specialized as a problem to obtain the assignment of the matrix T, which can achieve the target of Eq. (13). An algorithm Cost Matrix Generation shows the specific steps to calculate a cost matrix T through multiply the distribution C by distance matrix D.

4.2   Placement Algorithm for Reduce Tasks

In Algorithm 3, $getPartitionsList()$ method returns list partitions on the node i. Parameter D is the distance matrix. Method $Calculate()$ returns the value with the number of intermediate in the node n_i. And $Com()$ returns a minimum value of the array.

In Algorithm 4, the $getNodesList()$ method returns the list of nodes, and the $getCost()$ will obtain the cost of placing a jth reduce task on i^{th} node. The minimum value can be selected in the array by the method $Minimize()$.

Algorithm 3. Cost Matrix Generation

Require:
  n: the number of node;
  C: split-partition matrix;
  D: distance of resources.
Ensure:
  The cost matrix T.
1: $partitionList \leftarrow \varnothing$;
2: **for** each $i \in [0, n-1]$ **do**
3:     $PartitionList = getPartitionsList(C, i)$;
4:     **for** j in $partitionList$ **do**
5:         $tempC[j] \leftarrow Calculate(C, j, T)$;
6:     **end for**
7:     $tempT \leftarrow tempC \times getDist(D, j)$;
8:     $T \leftarrow Com(tempT)$;
9: **end for**
10: **return** T.

Algorithm 4. Reduce Task Placement

Require:
  r: the number of reduce;
  n: the number of node ;
  T: distance cost matrix.
Ensure:
  The placement queue R which is able to minimize the value of MC.
1: List $nodesList \leftarrow \varnothing$
2: **for** each $i \in [0, r-1]$ **do**
3:     $nodesList \leftarrow getNodesList(N, T)$;
4:     **for** j in nodes list **do**
5:         $tempSum[j] = getCost(T, i, j)$;
6:     **end for**
7:     $R[i] = Minimize(tempSum)$;
8: **end for**
9: **return** R.

---

5 Experimental Evaluation

5.1 Experiment Setting

Limited by the experimental conditions, theoretically our algorithms and models should get consistent results for large data sets. In the paper network congestion is not considered, and only for hops. The performance of different reduce placement algorithms are compared through two key indicators with the goal of measuring the impact of data locality. (1) Job execution time: This metric is used to measure execution time of the job. (2) Load balance: This metric is used to measure the Fos of the job. In our experiments, we chose the following reduce scheduling algorithms to compare with this algorithm.

NorP (Normal Placement). In the original Hadoop implements, the reduce tasks are launched according to random assignment and the resources utilization in the computing nodes.

Rang (Range Partition). This is a widely used algorithm of partition distribution. In this method, the intermediate $< key, value >$ tuples are sorted by key firstly, and then the tuples are assigned to reduce task according to this key range sequentially [15,20].

SARS (Self Adaptive Reduce Scheduling). An optimal reduce scheduling policy for reduce tasks start time in Hadoop platform. It can decide the start time points of each reduce tasks dynamically according to each job context, including the task completion time and the size of map output [21].

Our cluster consists of 20 physical machines installed the system of ubuntu 12.04 (KVM as the hypervisor) with 16 core 2.53 GHz Intel processors. Those machines are managed by CloudStack which is an open source cloud platform, having two racks and each rack containing 10 physical machines. A description of the various job types and the dataset sizes are shown follows: the input data of sort and grep are respectively 10 G and 10 G.

5.2   Performance Evaluation

Sampling Experiments. In this section, we set up ten groups of sampling experiments. According to the sampling rate, respectively, 25 %, 50 %, 75 %, and the real time(100 %), every group contains four sampling experiments. What the axis means is the Fos and execution time of the task in different sampling rate. As shown in Fig. 1, it is easy to learn that sampling 25 % of the map tasks can generate a sufficiently accurate approximation meanwhile with a relatively smaller time cost. Therefore, we set the default sampling rate as 25 % in this experiment, which can be changed by the user if necessary. To make the sampling threshold more accurate, we repeat our testing execution ten times completely through choosing the fixed map tasks with the same step interval as sample tasks. The Fig. 1 shows the accuracy of our sampling that would be the basis of for final results of the whole models.As above mentioned, the major motivation for CORP is to improve the data locality to diminish the cross-rack communication. Figure 2 compares the job executing time and Fos of these four strategies: Normal, Range partition, CORP, and SARS.

In the following experiments, to verify the advantages of CORP, we evaluate this model for Fos and job execution time using these common benchmarks: Sort and Grep [11].

Sort Benchmark Testing. Sort benchmark in Hadoop is usually used as workload testing because it is Map and Reduce-input heavy jobs. In these experiments, we generate 10 GB synthetic data sets following Zipf distributions with σ parameters varying from 0.1 to 1.2 to control the degree of the skew. The reason to choose Zipf to describe the frequency of the intermediate keys is that this distribution is very common in the data coming from the human society [22].

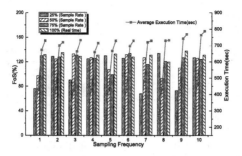

Fig. 1. The comparison experiment in various sampling rate

In the following experiments, the performance evaluations of the relevant algorithms are illustrated considering their job execution time in the change of the input data skew degree. In this paper, the CORP pays attention to achieve the best placement of a task,but SARS is mainly for the optimal starting time scheduling.These two combination being a time-space optimization of big data will bring significant results. From Fig. 2(a), we can conclude that CORP can improve the Fos obviously for different input data with various skew degree.

In more detail, as we can see from the Fig. 2(a), the value of Fos increases rapidly when the degree of the skew exceeds 0.7 for all the experimental algorithms, and from Fig. 2(b), the execution times of all algorithms also increased substantially once the degree of the skew reaches a certain threshold.

(a) load balance   (b) job execution time   (c) inner data traffic

Fig. 2. The performance evaluation for sort benchmark as the degree of data skew increases

The results in Fig. 2 shows that, through sampling to determine the placement of reduce tasks, CORP can get better data load balance, lower Fos and less transfer in the cross-rack comparing to other reduce scheduling algorithms. The reason that Range and CORP perform worse than NorP and SARS for execution time is that they need to pre-run the sample data which bring extra overhead.

Figure 3 represents the variation of job execution time while the data size rises from 4GB to 12GB with different skew degrees ($\sigma = 0.1$ and $\sigma = 1.2$). When the data set has a smaller skew degree (see Fig. 3(a)), SARS has the most time-efficient, and CORP performs only better than NorP due to the sample data pre-running costing some time.

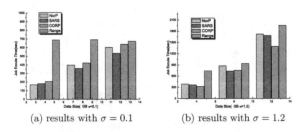

(a) results with $\sigma = 0.1$ (b) results with $\sigma = 1.2$

Fig. 3. The performance evaluation for sort benchmark as the size of input data increases.

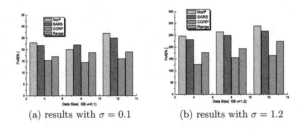

(a) results with $\sigma = 0.1$ (b) results with $\sigma = 1.2$

Fig. 4. Fos evaluation for sort benchmark as the size of input data increases.

Figure 4 illustrates the relationship between Fos and the reduce load: the smaller Fos, the better load balancing (keeping $\sigma = 1.2$). When the data set has a smaller skew degree and data size varies from 4 GB to 12 GB, we can observe that the Fos values of all methods increase slowly, but this indicator of CORP is always better than other compared algorithms.

Grep Benchmark Testing. Grep is a popular application for large scale data processing with heavy map-input. It searches some regular expressions through input text files and outputs the lines which contain the matched expressions. The data set we used is the full English Wikipedia archive with the total data size of 10 GB, which constitute the data processing jobs with heavy map tasks.

Since the behavior of Grep depends on how frequently the search expression appears in the input files, we tune the expression and make the input query percentage varies from 10 % to 100 %. Figure 5 shows the changes of job execution time Fos value and cross-rack transfer with the query percentage increases. Note that most current versions of Hadoop do not provide a suitable range partition for this application: their pre-run sampler can detect the input data but cannot handle applications where the intermediate data is in a different format from the input.

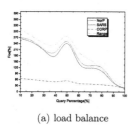
(a) load balance

(b) job execution time

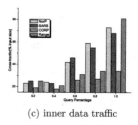
(c) inner data traffic

Fig. 5. The performance evaluation for Grep benchmark as the query percentage increases.

6 Conclusion

Because data sampling is an additional work for running jobs, it always brings the extra running times and degrades the overall system performance. As our previously proposed SARS is a time scheduling algorithm, CORP is an effective space scheduling algorithm of MapReduce tasks. Experiments verify that the data transmission among different nodes is optimized through reduce placement in CORP, and the whole execution time of a job can be decreased to offset the time spent on pre-running sampling. In a real system, the combination of CORP and SARS can improve the performance of MapReduce greatly.

References

1. Bourguiba, M., Haddadou, K., El Korbi, I., Pujolle, G.: Improving network i/o virtualization for cloud computing. IEEE Trans. Parallel Distrib. Syst. **25**(3), 673–681 (2014)
2. Ho, L.-Y., Wu, J.-J., Liu, P.: Optimal algorithms for cross-rackcommunication optimization in mapreduce framework. In: 2011 IEEE International Conference on Cloud Computing(CLOUD), pp. 420–427. IEEE (2011)
3. Isard, M., Prabhakaran, V., Currey, J., Wieder, U., Talwar, K., Goldberg, A.: Quincy: Fair scheduling for distributed computing clusters. In: Proceedings of the ACM SIGOPS 22nd Symposium on Operating Systemsprinciples, pp. 261–276. ACM (2009)
4. Ahmad, F., Lee, S., Thottethodi, M., Vijaykumar, T.: Mapreduce withcommunication overlap, technical report, Technical Report (2007)
5. Maheshwari, N., Nanduri, R., Varma, V.: Dynamic energy efficient data placement and cluster reconfiguration algorithm for mapreduce framework. Future Gener. Comput. Syst. **28**(1), 119–127 (2012)
6. Sandholm, T., Lai, K.: Dynamic proportional share scheduling in hadoop. In: Frachtenberg, E., Schwiegelshohn, U. (eds.) JSSPP 2010. LNCS, vol. 6253, pp. 110–131. Springer, Heidelberg (2010)
7. W. [EB/OL]. http://en.wikipedia.org/wiki/greedyalgorithm
8. Sleator, D.D., Tarjan, R.E.: Self-adjusting binary search trees. J. ACM (JACM) **32**(3), 652–686 (1985)

9. Kwon, Y., Balazinska, M., Howe, B., Rolia, J.: Skewtune: mitigating skew inmapre-duce applications. In: Proceedings of the 2012 ACM SIGMOD International Conference on Management of Data, pp. 25–36. ACM (2012)

10. Xu, Y., Kostamaa, P.: Efficient outer join data skew handling in parallel dbms. Proc. VLDB Endowment **2**(2), 1390–1396 (2009)

11. Palanisamy, B., Singh, A., Liu, L., Jain, B.: Purlieus: locality-awareresource allocation for mapreduce in a cloud. In: Proceedings of 2011International Conference for High Performance Computing, Networking, Storageand Analysis, p. 58. ACM (2011)

12. Ardizzoni, E., Bertossi, A., Pinotti, M.C., Ramaprasad, S., Rizzi, R., Shashanka, M.V., et al.: Optimal skewed data allocation on multiple channels with flat broadcast per channel. IEEE Trans. Comput. **54**(5), 558–572 (2005)

13. Acharya, S., Gibbons, P.B., Poosala, V.: Congressional samples forapproximate answering of group-by queries. ACM SIGMOD Rec. **29**(2), 487–498 (2000). ACM

14. Kwon, Y., Balazinska, M., Howe, B.: Skewtune: mitigating skew inmapreduce applications. In: 2012 ACM SIGMODInternational Conference on Management of Data, pp. 25–36. ACM (2012)

15. Chen, Q., Liu, C., Xiao, Z.: Libra: Lightweight data skew mitigation in mapreduce. IEEE Trans. Parallel Distrib. Syst. **99**, 1–14 (2014)

16. Lin, J.: The curse of zipf and limits to parallelization: A look at thestragglers problem in mapreduce. In: 7th Workshop on Large-Scale Distributed Systems for Information Retrieval, vol. 1 (2009)

17. Chen, Y., Wei, H.W., Wei, M.F., Chen, Y.J.: LaSA: A locality-aware scheduling algorithm forhadoop-mapreduce resource assignment. In: 2013 International Conference on Collaboration Technologies and Systems (CTS), pp. 342–346. IEEE (2013)

18. Hammoud, M., Sakr, M.F.: Locality-aware reduce task scheduling formapreduce. In: 2011 IEEE Third nternational Conference on Cloud Computing Technology and Science 29 2011-December 1 2011

19. Grover, R., Carey, M.J.: Extending map-reduce for efficient predicate-basedsampling. In: 2012 IEEE 28th InternationalConference on Data Engineering (ICDE), pp. 486–497. IEEE (2012)

20. Atta, F., Viglas, S.D., Niazi, S.: Sand joina skew handling joinalgorithm for google's mapreduce framework. In: 2011 IEEE 14th International Multitopic Conference, pp. 170–175. IEEE (2011)

21. Tang, Z., Jiang, L.G., Zhou, J.Q., Li, K.K., Li, K.K.: A self-adaptive scheduling algorithm for reduce start time. Future Gener. Comput. Syst. **43**, 51–60 (2015)

22. Gufler, B., Augsten, N., Reiser, A., Kemper, A.: Load balancing in mapreducebased on scalable cardinality estimates. In: 2012 IEEE 28th International Conference on Data Engineering (ICDE), pp. 522–533. IEEE (2012)

AAA: A Massive Data Acquisition Approach in Large-Scale System Monitoring

Runlin Zhou, Yanfei Lv[✉], Dongjin Fan, Hong Zhang, and Chunge Zhu

National Computer Network Emergency Response Technical Team/Coordination
Center of China, Beijing 100029, China
{zhourunlin,lyf,fdj,zhangh,zcg}@cert.org.cn

Abstract. The rapid development of information system proposes higher demand for monitoring. Usually we resort to a data acquisition system to collect variety of metrics from each device for real-time anomaly detection, alerting and analysis. It is a great challenge to realize real-time and reliable data collection and gathering in a data acquisition system for large-scale system. In this paper, we propose an Adaptive window Acquisition Algorithm (AAA) to support data acquisition on great amount of data sources. AAA can dynamically adjust its policy according to the number of data sources and the acquisition interval to achieve better performance. The algorithm has been applied to a large management system project. Experimental results show that with the help of dynamic adjusting mechanism, the proposed approach can provide reliable collection service for common data acquisition systems.

Keywords: System monitoring · Cloud computing · Acquisition service · Adaptive window

1 Introduction

The development of cloud computing and big data technologies make it possible to build large-scale information systems. A big information system often distributes across several data centers over a large network, which processes huge amount of data in order to meet the information needs in a timely manner.

To ensure information systems can provide reliable service, the demand of monitoring of such large-scale system becomes higher and higher, and thus many monitoring systems are proposed. The real-time data acquisition is one of the most important challenges in these systems. Although separated data acquisition systems can be built for different systems and work independently, with the increment of the number of equipment monitored, this mode is really inefficient and suffers several drawbacks. First, resources cannot be shared among data acquisition systems. Some systems may be idle while others are quite busy. Second, it is time-consuming to build data collection for new systems, which lead to the newly built systems cannot be monitored in time. In addition, building and maintaining so many data acquisition systems are time and money waste. Consequently, constructing a unified data collection platform and collection algorithm

© Springer International Publishing Switzerland 2016
J. Zhan et al. (Eds.): BPOE 2015, LNCS 9495, pp. 107–116, 2016.
DOI: 10.1007/978-3-319-29006-5_9

that fits for different kinds of monitoring applications will be a good choice for most information systems [4, 9].

In this paper, we proposed a massive data acquisition platform solution based on cloud computing. Data collection can be acquired on demand as a common service on this platform. The core algorithm in this platform is named Adaptive window Acquisition Algorithm (AAA for short). AAA approach serves as a generic collection component, which provides common data collection ability. We also deploy the approach in cloud to achieve better flexibility and scalability.

The general idea of AAA is: a mission submitted from cloud management platform is divided into small sub-task groups which are sorted according to time needed. A node will allocate each group a fixed resource to implement the tasks. Furthermore, the size of each group is adjusted dynamically according to time consumed by each data collection task to ensure the tasks can be completed within the stipulated time. If a node has reached the maximum limit of resources, the tasks on this node will be reassigned to other nodes [1, 10].

AAA algorithm has been applied to an IPMI network management system for a large-scale information system. Currently, the information system contains more than 2,000 physical machines distributed throughout the country. The collection component of IPMI network management system uses only two machines that can support up to 100,000 data collection tasks each day. The total collection supported is mainly limited by the execution time of a single IPMI collection task. In order to ensure the successful completion of all the tasks, IPMI network management system divides equipment monitored into groups, and uses AAA scheduling. Practical results show that AAA can ensure the implementation of each task even during peak hours.

Because to test AAA algorithm directly in the real environment may affect the stability of the information system, in this paper we realize an emulator to evaluate the effectiveness of AAA algorithm. Experimental results illustrate that the proposed algorithm demonstrates advantages in terms of usability and performance in comparison with random collection algorithm.

The remainder of this paper is organized as follows. We review the related work in Sect. 2. The AAA algorithm is presented in Sect. 3. Section 4 delivers the experimental results and finally we conclude the paper in Sect. 5.

2   Background and Related Work

2.1   Data Acquisition System

Monitoring system is an application system containing both software and hardware. The purpose of the system is to monitor the working status of the target information system, so as to ensure the information system in normal operating state. Monitoring system is the inevitable outcome of the development of computer system and developed along with the development of computer system. The quality of monitoring system significantly affects the early failure detection and treatment of the information system.

With fast development of computer technologies, the scale of information system increases rapidly with more complicated structure and more users supported. In order to cover a large scale system, monitoring system has to acquire and process large amount of working state data. The characteristics of the data are listed as follows:

(1) Large Volume. The data acquisition needs to cover heterogeneous integrated network with large scale and complex structure. In addition, the devices in the system may contain variety of equipment from different vendors. The amount of data needed to be collected by the monitoring system grows explosively with the increment of the scale and complexity of the information system.

(2) Uncertainty. The working status data of devices show obvious uncertainty characteristics. The uncertainty reflects not only on the data itself but also on the acquisition time of data. In order to obtain the accurate and timely information, it is necessary to preprocess and filter the data to be more comprehensive and more accurate.

Usually, a data acquisition system is adopted to perform collection and gathering of large amount of data from the target information system. A good data acquisition system can fit different kinds of network equipment, satisfy the restrict time requirement of monitoring. A data acquisition system should also be resource efficient, with good reliability and scalability.

In recent years, with the proposition of cloud computing, more and more systems are set up in the cloud. In this paper, the data acquisition system is also deployed in the cloud to facilitate the powerful processing capability of cloud computing and obtain better flexibility and scalability.

2.2   Related Work

Data acquisition system is an important part of the monitor system and responsible for the data extraction from the side of the working system to the event handlers for further processing. Therefore whether the data acquisition program can perform data collection accurately, stably and timely will directly affect the accuracy and stability of the monitoring system. In real networks, the data acquisition system shares network bandwidth with working system. In order to reduce the impact on the background traffic, the maximum bandwidth of data acquisition system should be limited to 5 percent of the whole bandwidth of network [2]. With the enlarging scale of the network and a sharp incensement in the number of equipment, the existing data acquisition system faces long acquisition cycle with heavy burden and too much network bandwidth occupation [11].

Many approaches have been proposed to design an effective and efficient data acquisition system to obtain realtime and reliable running state information at low cost of network communication. Dynamic polling algorithms based on cloud model or Markov chain are proposed in paper [3,6] respectively. By judging the magnitude of the fluctuation of the network, the management node adjusts the time interval of data acquisition to improve the efficiency of polling. Paper [8] proposed a rule based data acquisition system which establishes of a unified platform to provide effective data collection for different data sources. This kind of algorithm can effectively reduce the amount of communication data acquisition, but does not have a good scalability. When the number of nodes increases or network congestion occurs in the acquisition system, the burden on the management node increases accordingly, and acquisition time will be prolonged. Hierarchical and distributed network management technology [7,12] appears to solve the problems existing in the centralized data acquisition and improve the efficiency of the collection. Nevertheless, the random sequence acquisition strategy is applied in these approaches, so the efficiency of data acquisition declines at the time of network node failure or congestion [5].

3 Adaptive Window Acquisition Algorithm AAA

3.1 Definition of Collection Task

As illustrated in Fig. 1, assume that an information system contains a number of devices that need to be monitored. In this paper, these devices are named **collection targets**. In order to facilitate collection, some nodes are deployed which constantly send data request to the collection targets and return the results back. We called these nodes **collection nodes**. Each data collection on one target is named as a **collection task**. Usually, there are more than one collection nodes in an acquisition system and we use a node named **management node** to manage all these collection nodes.

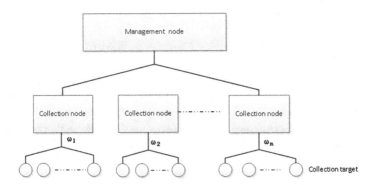

Fig. 1. Structure of AAA

The total collection target set that can be monitored by acquisition system is denoted by $\Psi = \{\omega_k | k = 1, 2, \cdots, l\}$, where ω_k stands for the set of targets whose working status is collected by the collection node k. For each device set ω, $\omega = \{r_j | j = 1, 2, \cdots, m\}$, r_j stands for the target set that needed to be collected in a single task. Each r_j will dynamically change with the resource on each node, noted as $r_j = \{s_i | i = 1, 2, \cdots, n\}$, where s_i represents one target that needs to be collected.

Suppose that all the collection targets in the subset of r_j have a unified collection task deadlines, denoted as T, that is, for a subset of targets r_j, the data collection task should be finished within time T. All the target subset r_j in set Ψ are scheduled in a parallel manner, the data collection tasks in r_j is scheduled using round robin algorithm.

3.2 AAA Algorithm

In this part, we present the detail of *Adaptive window Acquisition Algorithm*, AAA for short. The collection targets submitted by users are stored in the corresponding target set Ψ. In addition, the time of each collection task is recorded in collection node, depending on which, ω_k and r_j are determined.

The AAA algorithm includes two phases: (1) dynamic determination of group ω_k and (2) dynamic adjusting of window r_j. The first phase is executed on the management node, while the second runs on the collection nodes. The details of these two phases are listed in Algorithms 1 and 2 respectively.

In this algorithm, collection management node adopts the big round robin approach (Algorithm 1) to determine the data collection time ti on each target of all the collection nodes. According to the time ti, target groups of Ψ are assigned.

In the cloud node side (Algorithm 2) the algorithm determines the number of parallel programs dynamically depending on the number of r_j in ω. Subsequently the algorithm assigns one program to finish the collection target for the targets in one group.

After each collection cycle, the values of t_i on all targets are recalculated and sorted reversely with the last cycle, that is, if the values are sorted in ascending order in the last cycle, then theses values will be sorted in descending way in this cycle and vice versa. In this way, the data collection on all the targets can be fairly processed.

For the algorithm in Algorithm 2, if t_i of a target is greater than $1/2T$ and less than T, then the group r_j only contains this single target. In addition, if t_i is greater than T, then the algorithm exits with error, and the size of T will be readjusted.

The benefits of AAA algorithm are listed as follows.

- Adaptive to different kinds of data collection tasks. By adjusting the size of the target group dynamically, the value $\sum t_i$ in each group r_j is always less than $1/2T$, which guarantees collection frequency of each device group can be finished in a reasonable range.
- Covering all the targets monitored. By initializing the target group to the universal set, all the targets will be visited in a round-robin way. Furthermore, at the end of the each collection period or when a new device is added to the system, the whole algorithm will restart to ensure that the new device can be incorporated into monitor.
- Adaptive to the fluctuation of collection time. Sometimes, the collection time on certain devices may increase suddenly due to special causes such as unstable network. In this case, in order to ensure quality of service, the algorithm will terminate the data collection tasks of this group immediately, instead of waiting until the whole data collection cycle finishes.

Algorithm 1: AAA-1 algorithm

Input: the collection target device set Ψ, the number of nodes for collection node p

1 Initialize the collection time of each task $t_i = 0$;

2 Let $\Psi = \{\omega_k | k = 1\}$ that is the initial value of ω_1 is the whole set of targets which need to be collected;

3 Send the ω_1 as the collection target to each node in the cloud to estimate the collection time of each collection node to all the targets;

4 Each node execute the data collection task on the targets in ω_1 with concurrency of $\Psi/10$;

5 Record the time consumption of collection task on the corresponding target;
   /* This step is processed on collection node */;

6 Set $\Psi = \{\omega_k | k = 1, 2, \cdots, p\}$, where the size of ω_k is Ψ/p, which represents the resource allocated on the corresponding node;

7 Sort all the tasks according to collection time t_i in ascending order, the detail resource allocated on node k is to satisfy the collection tasks on top Ψ/p targets;

8 The resource allocated by node $k+1$ is top Ψ/p targets in set $\Psi - \sum \omega_k$ also sorted by t_i ;

9 After the determination of ω_k, the data collection tasks are sent to collection nodes and start to run;

10 **if** *The collection node number p changes* **then**

11   **if** *p increases* **then**

12   Execute step 4-5 on the newly arrival cloud node, and execute step 6-8 on the collection management node;

13   **else**   /* p decreases */

14   Execute step 6-8 on the collection management node directly;

15   **end**

16 **end**

17 **if** *The target set Ψ changes* **then**

18   **if** *Ψ increases* **then**

19   Treat ω as new targets, then execute step 4-5 on all the cloud nodes, subsequently execute step 6-8 on the collection management node;

20   **else**   /* Ψ increases */

21   Execute step 6-8 on the collection management node directly.

22   **end**

23 **end**

---

4   Experimental Evaluation

In this section, we implement a simulation environment to evaluate the effectiveness of AAA approach. The experimental results are illustrated in comparison with a random collection approach. Experimental study shows that our approach is superior to the random approach in both efficiency and reliability.

In the experiments, we run data collection task with 10 cycles using the random and AAA algorithm respectively, and record the collection time for each task, memory usage and the total time required. We set the total target devices $\Psi = 1000$, distributed evenly in five network segments. The number of collection

---

Algorithm 2: AAA-2 algorithm

Input: Given the collection target set ω, collection time limit T, collection
  frequency$1/T$, collection cycle Z

1 Sort the targets in r_j in ascending order according to collection time t_i obtained
  in step 5 in Algorithm 1 ;

2 Determine the size of r_j of each set, ensuring that Σt_i is not less than $1/2T$;

3 Generate j parallel tasks for the j device groups r_1 to r_j, and send to the cloud
  nodes;

4 For each target within the group r_j, execute the task sent by the step 3
  repeatedly until the task finished or the time limit T is reached;   /* This
  step is processed on cloud node */;

5 Repeat the step 3 and step 4, until the cycle Z finishes, or new target arrives;

6 Determine the new device resource set ω, and go back to step 1;

7 In each cycle, reverse the order in each group r_j;

---

cloud node P is set to 5, which are deployed in one segment with the same
hardware environment. One of the collection nodes serves as management node
as well.

4.1   Performance of AAA Algorithm

Figure 2 shows the time of each collection tasks and memory usage of each node.
The AAA algorithm can finish all the tasks in 25 s and the time of a single task is
in the 20–60 ms. However, in the random algorithm, only a small number of tasks
can be completed in 60 ms, the time of most tasks distribute around 120 ms. The
total time consumption is 55 s.

As can be seen from Fig. 2, AAA algorithm performs better than random
algorithm both in total time and the single task time. This is because the AAA
algorithm can learn the ability of each collection node on line and allocate suitable tasks to an efficient node.

4.2   Resource Consumption Evaluation

By fully utilizing of resources, the collection task can be ensured to be implemented in time, which is also achieved in a number of existing collection algorithm. Paper [11] pointed out that an excellent data collection algorithm should
not only ensure timely acquisition task scheduling, but also reduce the waste of
resources.

Figure 3 shows the memory usage of each collection node in the AAA algorithm and Random algorithm memory. As illustrated in Fig. 3(b) - (e), at the
beginning of AAA algorithm, node memory utilization has a quick burst. This is
because in this period of time, the nodes are estimating the collection time of all
the targets. Then the memory usage comes into a slow-growth period, in which

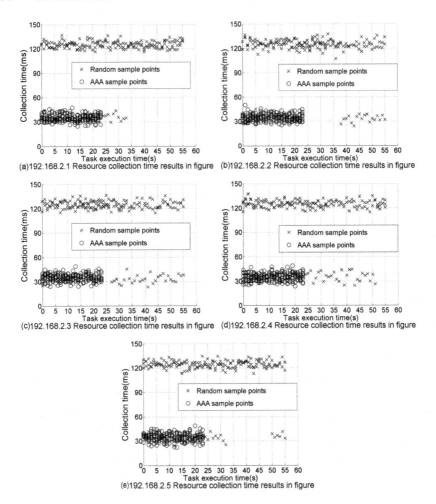

Fig. 2. Collection time comparison

collection tasks are performed collection nodes. In comparison, memory usage in random algorithm is growing more smoothly, since the collection resources are allocated in random way. Each node simply executes collection tasks in sequence order. Comparing the memory usage curves of two algorithms, AAA algorithm suffers higher memory usage during execution. Nevertheless, AAA algorithm only spends half the time of random algorithm. Consequently, the resource usage of AAA is more efficient for the same number of tasks.

In Fig. 3(a), the memory usage trends of AAA and random algorithm is the same with other subfigures, but the actual use of memory of random algorithm is always higher than AAA algorithm. This abnormal situation results from the presence of dirty data which should be eliminated.

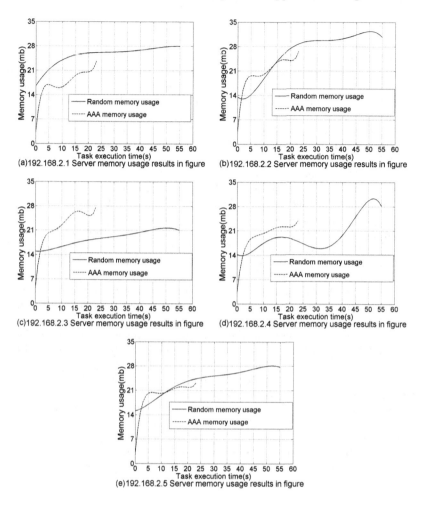

Fig. 3. Memory usage comparison

5 Conclusion and Future Work

This paper presents a massive resource acquisition mechanism: a resource reservation algorithm named AAA based on sliding window. With the help of rational resource allocation and dynamic window adjustment, AAA algorithm can not only satisfy all the data collection requests, but also increase the efficiency of data collection and enhance the reliability of system service.

AAA algorithm has been applied to an IPMI network management system for a large-scale information system. The future work includes further optimization of algorithms efficiency and deep analysis of algorithm performance in the real system.

References

1. Amazon elastic compute cloud (EC2). http://aws.amazon.com/ec2/
2. Boccardi, F., Huang, H.C.: Limited downlink network coordination in cellular networks. In: Proceedings of the IEEE 18th International Symposium on Personal, Indoor and Mobile Radio Communications, PIMRC, Athens, Greece, pp. 1–5 (2007)
3. Chris, T.K.N., Huang, H.: Linear precoding in cooperative MIMO cellular networks with limited coordination clusters. IEEE J. Sel. Areas Commun. **28**(9), 1446–1454 (2010)
4. Endo, P.T., de Almeida Palhares, A.V., Pereira, N.C.V.N., Gonçalves, G.E., Sadok, D., Kelner, J., Melander, B., Mångs, J.: Resource allocation for distributed cloud: concepts and research challenges. IEEE Netw. **25**(4), 42–46 (2011)
5. Funke, D., Brosig, F., Faber, M.: Towards truthful resource reservation in cloud computing. In: 6th International ICST Conference on Performance Evaluation Methodologies and Tools, pp. 253–262 (2012)
6. Hoydis, J., Kobayashi, M., Debbah, M.: On the optimal number of cooperative base stations in network MIMO systems. CoRR abs/1003.0332 (2010)
7. Liu, J., Wang, D.: An improved dynamic clustering algorithm for multi-user distributed antenna system. In: Proceedings of International Conference on Wireless Communications and Singal Processing, pp. 1–5 (2009)
8. Papadogiannis, A., Gesbert, D., Hardouin, E.: A dynamic clustering approach in wireless networks with multi-cell cooperative processing. In: Proceedings of IEEE International Conference on Communication, pp. 4033–4037 (2008)
9. Pawar, C.S., Wagh, R.B.: A review of resource allocation policies in cloud computing. World J. Sci. Technol. **2**(3), 165 (2012)
10. Sempolinski, P., Thain, D.: A comparison and critique of eucalyptus, opennebula and nimbus. In: Proceedings of the Second International Conference on Cloud Computing, pp. 417–426 (2010)
11. Venkatesan, S.: Coordinating base stations for greater uplink spectral efficiency in a cellular network. In: Proceedings of the IEEE 18th International Symposium on Personal, Indoor and Mobile Radio Communications, PIMRC Athens, Greece, pp. 1–5 (2007)
12. Zhou, S., Gong, J., Niu, Z., Jia, Y., Yang, P.: A decentralized framework for dynamic downlink base station cooperation. In: Proceedings of the Global Communications Conference, 2009. GLOBECOM, pp. 1–6 (2009)

Emerging Hardware

A Plugin-Based Approach to Exploit RDMA Benefits for Apache and Enterprise HDFS

Adithya Bhat[✉], Nusrat Sharmin Islam, Xiaoyi Lu, Md. Wasi-ur-Rahman,
Dipti Shankar, and Dhabaleswar K. (DK) Panda

Department of Computer Science and Engineering,
The Ohio State University, Columbus, USA
{bhata,islamn,luxi,rahmanmd,shankard,panda}@cse.ohio-state.edu

Abstract. Hadoop Distributed File System (HDFS) has been popularly
utilized by many Big Data processing frameworks as their underlying
storage engine, such as Hadoop MapReduce, HBase, Hive, and Spark.
This makes the performance of HDFS a primary concern in the Big Data
community. Recent studies have shown that HDFS cannot completely
exploit the performance benefits of RDMA-enabled high performance
interconnects like InfiniBand. To solve these performance issues, RDMA-
enabled HDFS designs have been proposed in the literature that show
better performance with RDMA-enabled networks. But these designs are
tightly integrated with the specific versions of the Apache Hadoop distribution, and cannot be used with other Hadoop distributions easily. In
this paper, we propose an efficient RDMA-based plugin for HDFS, which
can be easily integrated with various Hadoop distributions and versions
like Apache Hadoop 2.5 and 2.6, Hortonworks HDP, and Cloudera CDH.
Performance evaluations show that our plugin ensures the expected performance of up to 3.7x improvement in TestDFSIO write, associated
with the hybrid RDMA-enhanced design, to all these distributions. We
also demonstrate that our RDMA-based plugin can achieve up to 4.6x
improvement over Mellanox R4H (RDMA for HDFS) plugin.

1 Introduction

The Hadoop Distributed File System (HDFS) has taken a prominent role in the
data management and processing community and it has been becoming the first
choice to store very large datasets reliably on modern data center clusters. Many
companies, e.g. Yahoo!, Facebook, choose HDFS to manage and store petabyte or
exabyte-scale enterprise data. Because of its reliability and scalability, HDFS has
been utilized to support many Big Data processing frameworks, such as Hadoop
MapReduce [4], HBase [2], Hive [21], and Spark [24]. Such a wide adoption
has pushed the community to continuously improve the performance of HDFS
[19,20].

This research is supported in part by National Science Foundation grants #CNS-
1419123 and #IIS-1447804.

© Springer International Publishing Switzerland 2016
J. Zhan et al. (Eds.): BPOE 2015, LNCS 9495, pp. 119–132, 2016.
DOI: 10.1007/978-3-319-29006-5_10

To further improve the performance of HDFS operations, recent studies [9–12] have illustrated that HDFS communication and I/O performance can be significantly enhanced by taking advantage of the Remote Direct Memory Access (RDMA) feature on high performance interconnects, like InfiniBand and utilizing the heterogeneous storage devices available on modern clusters through efficient storage policies. RDMA-enabled HDFS designs [10,11] have been proposed and better performance results have been shown through these studies. However, these designs are tightly integrated with the specific versions of the Apache Hadoop distribution and cannot be integrated with other Hadoop distributions easily. This causes the unavailability of these new enhanced designs to the whole big data community, since many users are using different Hadoop versions and distributions. For example, Hortonworks HDP [7], Cloudera CDH are some of the most popular Hadoop distributions which have been deployed widely in many clusters. This motivates us to answer the question: *How can RDMA-enabled HDFS designs be utilized by different Hadoop distributions (e.g. Apache, HDP, and CDH) and versions without doing significant changes in the existing HDFS deployments?*

Also, in the Hadoop 2.x YARN architecture, a pluggable data shuffle mechanism has been introduced in the new MapReduce framework (MRv2), which allows developers and users to choose an alternative shuffle implementation on their demand, such as RDMA-based data shuffle proposed in [16,17,22]. This further motivates us to rethink the following broad design challenges:

1. Can we propose an RDMA-based plugin for HDFS, that can bring benefits of efficient hybrid, RDMA-enhanced HDFS designs, to different Hadoop distributions?
2. Can this plugin ensure similar performance benefits to Apache Hadoop as the existing enhanced designs of HDFS, such as [9–11]?
3. Can different enterprise Hadoop distributions, such as HDP [7], CDH also observe performance benefits for different benchmarks with RDMA plugin?
4. What is the possible performance improvement over other existing HDFS plugins such as Mellanox R4H [18]?

To explore the above challenges, this paper proposes an RDMA-based plugin for HDFS to leverage RDMA benefits across Apache and Enterprise Hadoop distributions. We provide an RDMA-based plugin which can be easily integrated with various Hadoop distributions and versions like Apache Hadoop 2.5.0 and 2.6.0, Hortonworks HDP 2.2 and Cloudera CDH 5.4.2. Extensive performance evaluations show that our plugin can ensure the expected performance associated with RDMA designs to all these distributions. Compared with default Hadoop packages running over IPoIB, our RDMA-based plugin can show as much improvement as the existing RDMA-enhanced designs in performances of different benchmarks, such as TestDFSIO, RandomWriter, Sort, TeraGen and TeraSort. We also demonstrate that our plugin performs significantly better than Mellanox R4H plugin [18].

The rest of the paper is organized as follows. Section 2 presents background for this paper. We give an overview of the existing RDMA-enhanced HDFS designs [9–11] in Sect. 3 and propose our RDMA-based plugin in Sect. 4. Section 5 describes our detailed evaluation. Section 6 provides other related studies currently existing in the literature. We conclude in Sect. 7 with possible future work.

2   Background

2.1   Apache Hadoop Distribution

Hadoop Distributed File System (HDFS) is the primary storage of a Hadoop cluster and the basis of many large-scale storage systems that is designed to span several commodity servers. Hadoop Distributed File System (HDFS) serves as efficient and reliable storage layer that provides high throughput access to application data.

Hadoop 2.x has made fundamental architectural changes to its storage platform, with the introduction of HDFS federation, NameNode high availability, Zookeeper-based fail-over controller, NFS support, and support for file system snapshots. While improvements to HDFS speed and efficiency are being added on an ongoing basis, different releases of the Apache Hadoop 2.x distribution have introduced significantly different versions of HDFS. For instance, Apache Hadoop 2.5.0, which is a minor release in the 2.x release includes some major features, such as specification for Hadoop compatible file system, support for POSIX-style file system attributes, while the Apache Hadoop 2.6.0 includes transparent encryption in HDFS along with a key management server, APIs for HDFS heterogeneous storage, work-preserving restarts of all YARN daemons, and several other improvements.

2.2   Enterprise Hadoop Distributions and Solutions

Hadoop has become the platform of choice for most businesses that are looking to harness the potential of Big Data. Enterprise versions like Hortonworks [7], Cloudera [3], etc., have been emerging to simplify working with Hadoop. Hortonworks Data Platform [7] (HDP) provides an enterprise ready data platform for a modern data architecture. This open Hadoop data platform incorporates new HDFS data storage innovations in version HDP 2.2, including heterogeneous storage, encryption, and some operational security enhancements. Cloudera Hadoop (CDH) [3] by Cloudera Inc., is a popular Hadoop distribution that provides a unified solution for batch processing, interactive SQL and search, along with role-based access controls, to streamline solving real business problems. Although the core components of these distributions are based on Apache Hadoop, the proprietary management and installation integrated into these distributions make it hard for users to consider trying out other HDFS enhancements. In addition to these integrated solutions, vendors such as Mellanox [13] have introduced solutions to accelerate HDFS, using RDMA-based plugin known as RDMA for HDFS (R4H) [18], that works alongside other HDFS communication layers.

3   Overview of RDMA-Enhanced HDFS Design

In the literature, there are many works that focus on improving the performance of HDFS operations on modern clusters. In [11,12], we propose an RDMA-based design of HDFS over InfiniBand that improves HDFS communication performance using pipelined and parallel replication schemes. In these works, the Java-based HDFS gets access to the native communication library via JNI that significantly improves the communication performance while keeping the existing HDFS architecture and APIs intact. SOR-HDFS [10] re-designs HDFS to maximize the overlapping among different stages of HDFS operations using a SEDA [23] based approach. This design proposes schemes to utilize the network and I/O bandwidth efficiently for maximal overlapping. These RDMA-enahnced designs work with different Apache Hadoop versions from 0.20.2 to 2.6.0 and guarantee performance benefits over default HDFS. In order to take advantage of the heterogeneous storage devices available on HPC clusters, we propose a new design [9], that accelerates HDFS I/O performance by intelligent placement of data via a hybrid approach. In this work [9], we propose enhanced data placement policies with hybrid replication and eviction/promotion of HDFS data based on the usage pattern. All these designs are done on top of Apache Hadoop codebase. In this study, we propose a corresponding plugin approach so that these high performing designs can be easily utilized by the Enterprise Hadoop versions. Our plugin incorporates the designs proposed in [10–12] and [9].

4   RDMA-Based Plugin Design for Apache and Enterprise HDFS

In this section, we discuss our design of the RDMA-based plugin for HDFS. The RDMA-based plugin for HDFS should provide same performance benefits as the existing RDMA enhanced HDFS designs [9–12]. The plugin approach should not introduce any overhead. Applying this plugin to different Hadoop distributions should involve minimal or no code change. Keeping these things in mind, RDMA-based plugin has client and server components. Client component gets loaded at the time the HDFS client JVM is launched to perform HDFS write operation. Server component is loaded when the DataNode is started.

Hadoop provides server and client side configuration to load user defined classes. By using such configuration we build RDMA plugin that can work across various Hadoop distributions. Figure 1(a) shows how the plugin fits into the existing HDFS structure. At the server side, there is RdmaDataXceiverServer that listens to RDMA connection and processes the blocks received. RdmaDataXceiverServer has to be loaded on every DataNode when it starts along with DataXceiverServer. This can be done by using the DataNode configuration parameter, dfs.datanode.plugins. In order for DataNode to load class using plugins parameter, it has to implement ServicePlugin interface and override start and stop methods provided by it. RdmaDataXceiverServer utilizes this methodology and implements ServicePlugin. Client side configuration parameter, fs.hdfs.impl is

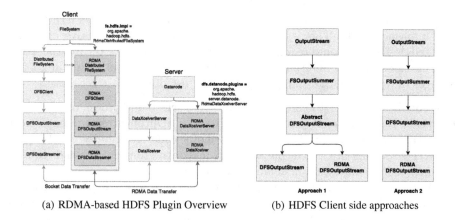

(a) RDMA-based HDFS Plugin Overview   (b) HDFS Client side approaches

Fig. 1. RDMA HDFS client plugin

used to load user designed filesystem instead of the default Hadoop distributed filesystem. We create a new filesystem class named RdmaDistributedFileSystem that extends DistributedFileSystem. We load RdmaDistributedFileSystem using client side parameter. Once RdmaDistributedFileSystem is loaded, it makes use of RdmaDFSClient to load RdmaDFSOutputStream. RdmaDFSOutputStream is one of the major component of the RDMA based HDFS plugin. For HDFS write operation, OutputStream class is the base abstract class that other classes need to extend if they wish to send data as a stream of bytes to some underlying link. The FSOutputSummer extends this class and adds additional functionality of generating the checksum for the data stream. DFSOutputStream extends FSOutputSummer. DFSOutputStream takes care of creating packets from the byte stream, adding these packets to the DataStreamer's queue and establishing the socket connection with the DataNode. DFSOutputStream has an inner class named DataStreamer that is responsible for creating a new block, sending the block as a sequence of packets to the DataNode and waiting for acknowledgments from the DataNode for the sent packets. For HDFS write operation, there are a set of common procedures that need to be followed to add the data into HDFS. The procedures are: (1) get the data from byte stream and convert them into packets, (2) contact the NameNode and get the block name and set of DataNodes, and (3) send a notification to the NameNode once the file is written completely. To design RdmaDFSOutputStream, we considered two approaches as shown in Fig. 1(b).

Approach 1: Figure 1(b) shows the first approach on the left-hand side. Here we create an abstract class AbstractDFSOutputStream which extends FSOutputSummer. This abstract class would contain implementation of common methods to communicate with the NameNode. Using this abstract class we can implement RdmaDFSOutputStream that can make use of common communication methods defined in the abstract class and can override methods that require RDMA specific implementation. This is a good approach from the perspective of object oriented design but would require code reorganization to move the existing common methods from DFSOutputStream to AbstractDFSOutputStream and make changes in default DFSOutputStream to make use of this abstract class.

Approach 2: Figure 1(b) shows the second and selected approach on the right-hand side. Here we directly extend the existing DFSOutputStream to implement RdmaDFSOutputStream. This approach allows code re-usability by changing the access specifier in DFSOutputStream from private to protected and requires minimal code modification in default Hadoop.

We package these RDMA specific plugin files into a distributable jar along with native libraries which are verbs-level communication. We need to change the access specifier in DFSOutputStream from private to protected that goes as a patch. So the plugin involves jar, native library, patch and a shell script to install this plugin. This script applies the patch to appropriate source files based on the version of the Hadoop distribution. RDMA specific files are bundled as a jar that needs to be added as a dependency in pom.xml file of Maven build system. This jar also needs to be copied to the Hadoop classpath. These will be taken care of by our provided shell script.

Figure 2 shows the RDMA plugin features. The RDMA plugin incorporates RDMA-based HDFS write, RDMA-based replication, RDMA-based parallel replication, SEDA designs proposed and Triple-H features in [9–12]. Figure 2 also depicts the evaluation methodology. RDMA plugin along with Triple-H [9] design is incorporated into Apache Hadoop 2.6, HDP 2.2 and CDH 5.4.2 code base which are shown as Apache-2.6-TripleH-RDMAPlugin, HDP-2.2-TripleH-RDMAPlugin and CDH-5.4.2-TripleH-RDMAPlugin legends in Sect. 5. Evaluation of Triple-H with RDMA in integrated mode is indicated by Apache-2.6-TripleH-RDMA. For Apache Hadoop 2.5, we apply the RDMA plugin without Triple-H design as it is not available for this version and this is shown as Apache2.5-SORHDFS-RDMAPlugin legend in Sect. 5. Figure 2(b) shows the evaluation methodology for IPoIB where we have used Apache Hadoop 2.6, 2.5, HDP 2.2 and CDH 5.4.2. These are addressed as Apache-2.6-IPoIB, Apache-2.5-IPoIB, HDP-2.2-IPoIB and CDH-5.4.2-IPoIB legends in Sect. 5.

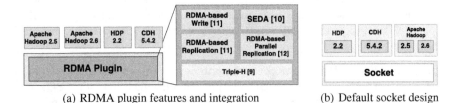

(a) RDMA plugin features and integration   (b) Default socket design

Fig. 2. Implementation and deployment

5   Performance Evaluation

In this section, we present a detailed performance evaluation of our RDMA-based HDFS plugin. We apply our plugin over Apache and Enterprise HDFS distributions. We use Apache Hadoop 2.5, Apache Hadoop 2.6, HDP 2.2 and CDH 5.4.2.

In this study, we do the following experiments:

(1) Evaluation with Apache Hadoop Distributions (2.5.0 and 2.6.0)
(2) Evaluation with Enterprise Hadoop Distributions and Plugin (HDP, CDH, and R4H)

5.1   Experimental Setup

(1) Intel Westmere Cluster (Cluster A): This cluster has Intel Westmere series processors using Xeon Dual quad-core processor nodes operating at 2.67 GHz with 12 GB RAM, 6 GB RAM Disk, and 160GB HDD. Each node has MT26428 QDR ConnectX HCAs (32 Gbps data rate) with PCI-Ex Gen2 interfaces and each node runs Red Hat Enterprise Linux Server release 6.1.

(2) Intel Westmere Cluster with Larger Memory (Cluster B): This cluster has Xeon Dual quad-core processor nodes operating at 2.67 GHz. Each node is equipped with 24 GB RAM and two 1TB HDDs. Four of the nodes have 300GB OCZ VeloDrive PCIe SSD. Each node is also equipped with MT26428 QDR ConnectX HCAs (32 Gbps data rate) with PCI-Ex Gen2 interfaces, and are interconnected with a Mellanox QDR switch. Each node runs Red Hat Enterprise Linux Server release 6.1.

In all our experiments, we use four DataNodes and one NameNode. The HDFS block size is 128 MB and replication factor is three. All the following experiments are run on Cluster B, unless stated otherwise.

5.2   Evaluation with Apache Hadoop Distributions

In this section, we evaluate the RDMA-based plugin using different Hadoop benchmarks over Apache Hadoop distributions. We use two versions of Apache Hadoop: Hadoop 2.6 and Hadoop 2.5.

Apache Hadoop 2.6: The proposed RDMA plugin incorporates our recent design on efficient data placement with heterogeneous storage devices, Triple-H [9] (based on Apache Hadoop 2.6), RDMA-based communication [11] and enhanced overlapping [10]. We apply the RDMA plugin to Apache Hadoop 2.6 codebase and compare the performances with those of IPoIB. For this set of experiments, we use one RAM Disk, one SSD, and one HDD per DataNode as HDFS data directories for both default HDFS and our designs.

In the graphs, default HDFS running over IPoIB is indicated by Apache-2.6-IPoIB, Triple-H design without the plugin approach is indicated by Apache-2.6-TripleH-RDMA and Triple-H design with RDMA-enhanced plugin is shown by Apache-2.6-TripleH-RDMAPlugin.

Figure 3 shows the performance of TestDFSIO Write test. As observed from the figure, our plugin (Apache-2.6-TripleH-RDMAPlugin) does not incur any significant overhead compared to Apache-2.6-TripleH-RDMA and is able to offer similar performance benefits (48 % reduction in latency and 3x improvement in throughput) like that of Apache-2.6-TripleH-RDMA.

Figure 4 shows the performance comparison of different MapReduce benchmarks. Figure 4(a) and (b) present performance comparison of TeraGen and RandomWriter respectively, for Apache-2.6-IPoIB, Apache-2.6-TripleH-RDMA, and Apache-2.6-TripleH-RDMAPlugin. The figure shows that, with the plugin, we are able to achieve similar performance benefits (27 % for TeraGen and 31 % for RandomWriter) over IPoIB as observed for Apache-2.6-TripleH-RDMA.

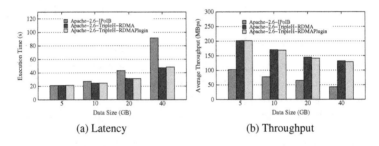

(a) Latency   (b) Throughput

Fig. 3. Evaluation of HDFS plugin with Hadoop 2.6 using TestDFSIO

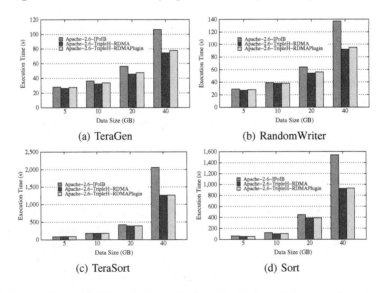

(a) TeraGen   (b) RandomWriter

(c) TeraSort   (d) Sort

Fig. 4. Evaluation of HDFS plugin with Apache Hadoop 2.6 using data generation benchmarks

The Triple-H design [9], along with the RDMA-enhanced designs [10,11], incorporated in the plugin, improve the I/O and communication performance, and this in turn leads to lower execution times for both of these benchmarks. We also evaluate our plugin using the TeraSort and Sort benchmarks. Figure 4(c) and (d) show the results of these experiments. As observed from the figures, the Triple-H design, as well as the RDMA-based communication incorporated in the plugin, ensure similar performance gains (up to 39 % and 40 %, respectively) over IPoIB for TeraSort and Sort as Apache-2.6-TripleH-RDMA.

Apache Hadoop 2.5: In this section, we evaluate our plugin with Apache Hadoop 2.5. Triple-H design is not available for this Hadoop version. Therefore, we evaluate default HDFS running over IPoIB, and compare it with RDMA-enhanced HDFS [10] used as a plugin. In this set of graphs, default HDFS running over IPoIB is indicated by Apache-2.5-IPoIB and our plugin by Apache-2.5-SORHDFS-RDMAPlugin.

Figure 5 shows the results of our evaluations with the TestDFSIO write benchmark. As observed from this figure, our plugin can ensure the same level of performance as SOR-HDFS [10], which is up to 27 % higher than that of default HDFS over IPoIB in terms of throughput for 40 GB data size. For latency on the same data size, we observe 18 % reduction with our plugin, which is also similar to SOR-HDFS [10].

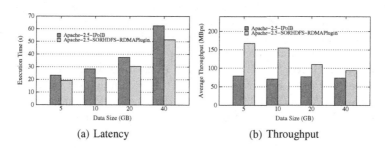

(a) Latency   (b) Throughput

Fig. 5. Evaluation of SOR-HDFS RDMA plugin with Apache Hadoop 2.5 using TestDFSIO

5.3   Evaluation with Enterprise Hadoop Distributions

In this section, we evaluate the RDMA-based plugin using different Hadoop benchmarks over Enterprise Hadoop distributions: HDP 2.2 and CDH 5.4.2. We apply the RDMA plugin to HDP 2.2 and CDH 5.4.2 codebases and compare the performances with those of the default ones.

Figure 6 shows the performance of the TestDFSIO Write test. As observed from the figure, our plugin (HDP-2.2-TripleH-RDMAPlugin) is able to offer similar performance gains as shown in [9] for Apache distribution, to HDP 2.2. For example, with HDP-2.2-TripleH-RDMAPlugin, we observe 63 % performance benefit compared to HDP-2.2-IPoIB in terms of latency. In terms of throughput,

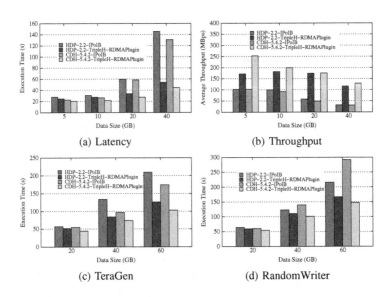

Fig. 6. Evaluation of Triple-H RDMA plugin with HDP-2.2 using TestDFSIO

the benefit is 3.7x. The Triple-H-RDMA plugin brings similar benefits for the CDH distribution as well through the enhanced designs [9–11] incorporated in it.

In Fig. 6(c) and (d), we present the performance comparisons for data generation benchmarks, TeraGen and RandomWriter respectively, for HDP and CDH distributions. The figure shows that, with the plugin applied to HDP 2.2, we are able to achieve similar performance benefits (37 % for TeraGen and 23 % for RandomWriter) over IPoIB, as observed in [9] for Apache distribution. The benefit comes from the improvement in I/O and communication performance through Triple-H [9] and RDMA-enhanced designs [10,11]. Similarly, the benefits observed for CDH distribution are 41 % for TeraGen and 49 % for RandomWriter.

We also evaluate our plugin using the TeraSort and Sort benchmarks for HDP 2.2. Figure 7(a) and (b) show the results of these experiments. As observed from

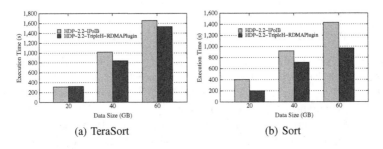

Fig. 7. Evaluation of Triple-H RDMA plugin with HDP-2.2 using different benchmarks

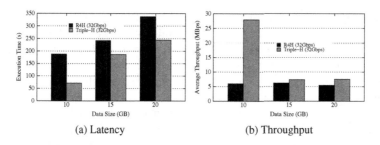

(a) Latency   (b) Throughput

Fig. 8. Performance comparison between Triple-H RDMA plugin and R4H (Cluster A)

the figures, the Triple-H design, as well as the RDMA-based communication incorporated in the plugin, ensure performance gains of up to 18 % and 33 % over IPoIB for TeraSort and Sort, respectively.

Figure 8 shows the TestDFSIO latency and throughput performance between R4H applied to HDP 2.2 and HDP-2.2-TripleH-RDMAPlugin. We use four nodes on cluster A for this experiment as R4H requires Mellanox OFED and cluster B has OpenFabrics OFED. As observed from the figures, for HDP-2.2-TripleH-RDMAPlugin, we see up to 4.6x improvement in throughput compared to R4H for 10 GB data size. As the data size increases, throughput becomes bounded by disk. Even then the enhanced I/O and overlapping designs of Triple-H and SOR-HDFS that are incorporated in the plugin lead to 38 % improvement in throughput for 20 GB data size. The improvement in latency is 28 % for 20 GB and 2.6x for 10 GB data size.

6   Related Work

With the Hadoop 2.x framework, Next Generation MapReduce (MRv2) has introduced pluggable shuffle and sort capabilities to the Hadoop community, allowing users to replace the built-in shuffle and sort logic with alternate implementations, such as RDMA-based data shuffling for MapReduce, proposed in [16,17], or custom sort algorithms that enable hash aggregation and Limit-N query. Along similar lines, we propose an RDMA-based plugin for the Hadoop Distributed File System (HDFS), that can take advantage of various RDMA-enhanced and hybrid enhancements proposed for HDFS, to different Hadoop distributions, with minimal effort.

There has been immense advancement in the Hadoop world, with the extensive research [20] carried out to advance the Hadoop Distributed File System (HDFS). Recent research works have proposed RDMA-enhanced HDFS designs [11,12] to improve HDFS performance using pipelined and parallel replication schemes. Similarly, a SEDA (Staged Event-Driven Architecture) [23] based approach to re-design HDFS architecture, namely SOR-HDFS [10], that maximizes the possible overlapping during different stages, such as processing, communication, and I/O, to eliminate significant bottlenecks in HDFS. In addition to these, the R4H (RDMA for HDFS) plugin [18] introduced by Mellanox [13], accelerates HDFS using RDMA, and demonstrates an improvement of 40–50 % in MapReduce CPU time.

Recent studies [1,5] have shown that the read throughput of HDFS can be improved significantly by employing mechanisms to caching data in memory. In a similar context, [8] leverages Memcached as a caching layer to improve both HDFS read and write performance. To exploit the heterogeneous storage architecture, that is available in most high performance cluster today, a hybrid approach was proposed in [9]. Similarly, [14,15] also attempt to incorporate heterogeneous storage media such as SSD into HDFS. To better understand these advancements, and leverage them across different Hadoop distributions like Apache, Hortonworks, Cloudera etc., we believe that the RDMA-based plugin proposed in this paper, will definitely be beneficial.

7   Conclusion

In this paper, we propose an RDMA-based plugin for Hadoop Distributed File System (HDFS), to leverage the benefits of RDMA across different Hadoop distributions, including Apache and Enterprise. The performance benefits shown by existing RDMA enhanced HDFS designs [9–12] can be achieved in this proposed RDMA-based plugin for HDFS. This flexible technique provides a smart and efficient way to intergrate fast data transmission functionalities from existing hybrid RDMA-enhanced HDFS designs, while adapting the general plugin-based approach in HDFS, in order to bring the network-level benefits to the end applications. We implement our RDMA-based plugin that includes the RDMA enhancements existing in the literature [10,11] and apply on top of different Hadoop distributions.

Our experimental results demonstrate that our proposed RDMA-based HDFS plugin incurs no extra overhead in terms of performance for different benchmarks. We observe up to 3.7x improvement in TestDFSIO write throughput, and up to 48 % improvement in latency, as compared to different Hadoop distributions running over IPoIB. We also demonstrate that our plugin can achieve up to 4.6x improvement in TestDFSIO write throughput, and 62 % improvement in TestDFSIO write latency, as compared to Mellanox R4H plugin. We have currently made RDMA-based HDFS plugin with Triple-H design [9] publicly available as a part of the HiBD project [6], for Apache Hadoop and HDP. In the future, we plan to make the RDMA-based plugin available for other popular Hadoop distributions such as CDH. We plan to undertake detailed studies to assess the benefits of using the proposed plugin for real-world Hadoop applications and database workloads like YCSB using HBase.

References

1. Ananthanarayanan, G., Ghodsi, A., Wang, A., Borthakur, D., Kandula, S., Shenker, S., Stoica, I.: PACMan: Coordinated memory caching for parallel jobs. In: Proceedings of the 9th USENIX Conference on Networked Systems Design and Implementation. NSDI 2012, San Jose, CA (2012)
2. Apache HBase. http://hbase.apache.org/

3. Cloudera Hadoop Distribution: http://cloudera.com/
4. Dean, J., Ghemawat, S.: MapReduce: Simplified data processing on large clusters. In: OSDI. Boston, MA (2004)
5. Foundation, A.S.: Centralized Cache Management in HDFS. http://hadoop.apache.org/docs/r2.3.0/hadoop-project-dist/hadoop-hdfs/CentralizedCacheManagement.html
6. Data, High-Performance Big (HiBD). http://hibd.cse.ohio-state.edu
7. Hortonworks: We do Hadoop. Enabling the Data-First Enterprise. http://hortonworks.com/
8. Islam, N.S., Lu, X., Rahman, M.W., Rajachandrasekar, R., Panda, D.K.: In-memory I/O and replication for HDFS with memcached: Early experiences. In: 2014 IEEE International Conference on Big Data (IEEE BigData). Washington DC (2014)
9. Islam, N.S., Lu, X., Rahman, M.W., Shankar, D., Panda, D.K.: Triple-H: A hybrid approach to accelerate HDFS on HPC clusters with heterogeneous storage architecture. In: 15th IEEE/ACM International Symposium on Cluster, Cloud and Grid Computing. China (2015)
10. Islam, N.S., Lu, X., Rahman, M.W., Panda, D.K.: SOR-HDFS: A SEDA-based approach to maximize overlapping in RDMA-enhanced HDFS. In: The Proceedings of The 23rd International ACM Symposium on High-Performance Parallel and Distributed Computing (HPDC). Canada (2014)
11. Islam, N.S., Rahman, M.W., Jose, J., Rajachandrasekar, R., Wang, H., Subramoni, H., Murthy, C., Panda, D.K.: High performance RDMA-based design of HDFS over infiniBand. In: The Proceedings of The International Conference for High Performance Computing, Networking, Storage and Analysis (SC). Salt Lake City (2012)
12. Islam, N.S., Lu, X., Rahman, M.W., Panda, D.K.: Can parallel replication benefit hadoop distributed file system for high performance interconnects? In: The Proceedings of IEEE 21st Annual Symposium on High-Performance Interconnects (HOTI). San Jose, CA (2013)
13. Mellanox. http://www.mellanox.com
14. Anwar, R.K., Butt, A.A.: hatS: A heterogeneity-aware tiered storage for hadoop. In: 14th IEEE/ACM International Symposium on Cluster, Cloud and Grid Computing (CCGRID) (2014)
15. R.K., Iqbal, S., Butt, A.: VENU: Orchestrating SSDs in hadoop storage. In: 2014 IEEE International Conference on Big Data (IEEE BigData) (2014)
16. Rahman, M.W., Islam, N.S., Lu, X., Jose, J., Subramoni, H., Wang, H., Panda, D.K.: High-performance RDMA-based design of hadoop mapreduce over infiniBand. In: HPDIC, in conjunction with IPDPS. Boston, MA (2013)
17. Rahman, M.W., Lu, X., Islam, N.S., Panda, D.K.: HOMR: A hybrid approach to exploit maximum overlapping in mapreduce over high performance interconnects. In: ICS. Munich, Germany (2014)
18. RDMA for HDFS (R4H). https://github.com/Mellanox/R4H
19. Shafer, J., Rixner, S., Cox, A.: The hadoop distributed filesystem: Balancing portability and performance. In: 2010 IEEE International Symposium on Performance Analysis of Systems Software (ISPASS), pp. 122–133, March 2010
20. Shvachko, K.: HDFS Scalability: The Limits to Growth (2010)
21. The Apache Software Foundation: The Apache Hive. http://hive.apache.org/
22. Wang, Y., Que, X., Yu, W., Goldenberg, D., Sehgal, D.: Hadoop acceleration through network levitated merge. In: SC (2011)

23. Welsh, M., Culler, D., Brewer, E.: SEDA: An architecture for well-conditioned, scalable internet services. In: Proceedings of the 18th ACM Symposium on Operating Systems Principles (SOSP). Banff, Alberta, Canada (2001)
24. Zaharia, M., Chowdhury, M., Franklin, M.J., Shenker, S., Stoica, I.: Spark: Cluster computing with working sets. In: Proceedings of the 2Nd USENIX Conference on Hot Topics in Cloud Computing. HotCloud 2010, Boston, MA (2010)

Stream-Based Lossless Data Compression Hardware Using Adaptive Frequency Table Management

Shinichi Yamagiwa[1]([✉]), Koichi Marumo[1], and Hiroshi Sakamoto[2]

[1] Faculty of Engineering, Information and Systems/Department of Computer Science, University of Tsukuba, 1-1-1 Tennodai, Tsukuba, Ibaraki, Japan
yamagiwa@cs.tsukuba.ac.jp, marumo@padc.cs.tsukuba.ac.jp
[2] Graduate School of Computer Science and Systems Engineering, Kyushu Institute of Technology, 680-4 Kawazu, Iizuka-shi, Fukuoka, Japan
hiroshi@ai.kyutech.ac.jp

Abstract. In order to treat BigData efficiently, the communication speed of the inter or the intra data path equipped on high performance computing systems that needs to treat BigData management has been reaching to very high speed. In spite of fast increasing of the BigData, the implementation of the data communication path has become complex due to the electrical difficulties such as noises, crosstalks and reflections of the high speed data connection via a single cupper-based physical wire. This paper proposes a novel hardware solution to implement it by applying a stream-based data compression algorithm called the LCA-DLT. The compression algorithm is able to treat continuous data stream without exchanging the symbol lookup table among the compressor and the decompressor. The algorithm includes a dynamic frequency management of data patterns. The management is implemented by a dynamic histogram creation optimized for hardware implementation. When the dedicated communication protocol is combined with the LCA-DLT, it supports remote data migration among the computing systems. This paper describes the algorithm design and its hardware implementation of the LCA-DLT, and also shows the compression performance including the required hardware resources.

Keywords: Lossless data compression · Communication protocol · Dynamic histogram creation

1 Introduction

BigData management performed by the latest high performance computing system must contain indispensable high speed data paths such as the intra-connection between the processor and the memory and/or the inter-connection among the multiple processors. The processors communicate with the other ones via peripheral busses such as the PCI Express bus in GHz order. The peripheral

© Springer International Publishing Switzerland 2016
J. Zhan et al. (Eds.): BPOE 2015, LNCS 9495, pp. 133–146, 2016.
DOI: 10.1007/978-3-319-29006-5_11

devices such as network adapters communicate each other via cupper or optical wires in tens of GHz order. These high speed communication data busses in the future will reach to very higher speed until the technical limitation for implementation. To avoid such emergent technical limitation, we are trying to parallelize the data path to multiple connections such as 256 bits in a processor bus, 16 lanes in the PCI bus and multiple wires in the network. However, these technological trials are not radical solution in order to implement scalable data communication path for BigData management.

To improve the future implementation limitation, we focus on data compression on the high speed data path. There are two ways for the data compression in the path. One is the software-based compression in which a compression algorithm is implemented on the lower layer in the communication path. Although we had implemented it in the device driver level of Ethernet, it has become just overhead to impede the communication [14]. Therefore, another way is indispensable to provide data compression mechanism on the communication path based on hardware implementation. Hence, the hardware must provide low latency and stream-based compression/decompression data flow.

The conventional data compression algorithms such as Huffman encoding [13] and LZW [15,16] perform data encoding creating *symbol lookup table*. The frequent data patterns are replaced by the compressed symbol in the table. However when we consider the hardware-based implementation, we can find difficulties: (1) unpredictable processing time for compression/decompression because the data length is not deterministic, (2) unpredictable required memory quantity during the compression because the number of entries in the symbol lookup table to count the frequency of data patterns is not deterministic and (3) blocking decompression is performed until the compression process is finished. These problems disturb continuous compression to perform streaming data propagation in the communication path. Therefore, it is necessary to invent a new compression algorithm.

This paper proposes a novel data compression method for high speed communication path based on hardware implementation. The proposed compression/decompression algorithm called the LCA-DLT manages the frequency of symbol appearance on a hardware-based dynamic histogram that can work on a limited number of memory resources. Moreover, it does not send the symbol lookup table to the decompressor. The compressor transmits only the data stream in pipeline manner, and then the decompressor can begin to decode the compressed data stream one after another to the originals before receiving the entire compressed data.

The next section will describe the backgrounds and definitions regarding the communication methods in the recent computing platforms and the conventional data compression mechanism. After the discussion for the needs of compression on the high speed communication path, this paper will propose a novel algorithm for hardware-based compression on communication. Section 3 will show design and implementation of the compression method. Section 4 will show performance evaluation in order to validate the effect of the data compression on communication. Finally, Sect. 5 will conclude the paper.

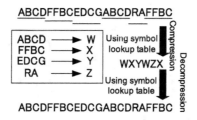

Fig. 1. The conventional compression algorithm.

2   Backgrounds and Definitions

2.1   Data Compression on Communication

To increase bandwidth of high speed communication data path, we have three major methods. The first one is to increase the bus frequency. The second one is to prepare parallel paths such as multiple cupper wires. These solutions would bring implementation difficulties because the higher frequency brings electrical problems regarding the impedance of the communication paths or the signal propagation delays among multiple wires. The third one is to decrease data quantity in the communication path. This is a promising technique to increase the bandwidth. However, reducing data in the application level forces modifications of algorithms. This is not suitable for the existing application programs. Thus the data compression in the communication path is the suitable solution for adaptively increasing the bandwidth.

When we apply data compression in the communication path, we need to consider an effective layer in a system where we apply a data compression. The highest layer in a computing platform is application. However, introducing the data compression lets us to modify the application. Therefore, it is necessary to introduce it in the lower layer than the application. It can be implemented on OS or hardware layer.

Here, let us see a trial of data compression in the link layer of Ethernet studied by the author [14]. In the trial, data compression algorithm (LZO which is an optimized Lempel-Ziv algorithm) was embedded in the device driver of Ethernet NIC just before the Ethernet packet creation. While the network adapter works at 100 Mbps based on 1500 byte packet, it effectively compressed the communication data from/to any programs among remote computing platforms. It achieved about 400 Mbps in a peer-to-peer application. However, the performance degraded if the adapter worked as a Gbit Ethernet. It achieved only 800 Mbps because the software-based calculation time of the compression algorithm had become redundant overhead during the communication. According to the experimental trial above, we need to implement a data compression algorithm in hardware layer with small latency for data flow as possible.

2.2   Data Compression Algorithms

Well-known data compression algorithms such as LZW [15,16] perform the steps as shown in Fig. 1. When a data stream is inputted to the compressor, it finds

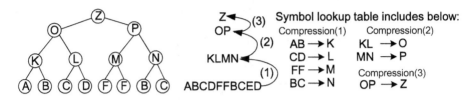

Fig. 2. An example of succinct data structre using binary tree.

the longest frequent patterns and registers those to a *symbol lookup table*. Here, the unit of the data stream is called as *symbol*. In the figure, the symbol translation pattern "ABCD→W" is registered to the symbol lookup table at the beginning. When the pattern is replaced to a symbol that never appears in the stream, the pattern is compressed. Therefore, we use "W" for the compressed symbol for the pattern and then the symbol is outputted. As processing the rest of the data stream till the end of the stream, the compressed data stream becomes the compressed one shown in the middle of the figure, and then a symbol lookup table is completely generated. The decompressor side decodes the compressed data stream using the lookup table with matching the compressed symbol to the original pattern. Therefore the decompressor needs to receive the lookup table and translates the compressed data stream to the original one by using it.

Assume that the data compression algorithm above is applied to a communication data path that needs to send a data stream from a computing platform to another. The compressor must scatter the stream data to several chunks of data because the steps need to synchronize the lookup table in the end of the data chunk. This means that the decompressor must stay in idle during the compression process. Because the serialization of the communication degrades the performance, the compression and the decompression must be pipelined based on a fine grain data unit. However, it is impossible to implement the data pipeline due to unavoidable receiving of the lookup table in the decompressor side. Thus, we need to develop a new compression algorithm without exchanging the table between compressor and decompressor. If the exchange is eliminated from the communication, the compressed data stream can be propagated in pipeline manner between the compressor and the decompressor. It can exploit the peak performance of the physical media and then the effective bandwidth will become higher than the physical peak performance in a lower speed if the original data is compressed. This will provide easy implementation availability for a high speed communication data path.

In the research field of the lossless data compression based on the succinct data structure, we can find several effective methods. Let us consider if we can implement those in streaming hardware compressor/decompressor or not. First, the dictionary-based method is popular since the LZW compression. The methods [2,6,8,12] prepare dictionary that is similar to the symbol lookup table, and compress input data by replacing the frequent symbol patters registered in the dictionary to an unused symbol. This mechanism needs to exchange the dictionary from the compressor to the decompressor and also needs the unpredictable size of working memory. The other methods [3,4] provide a compression mechanism adjusting entropy of the input data. This method is too hard to implement on hardware because it needs unexpected memory size during the compression operation. On the other hand, the method called LCA [10] seems to fit into hardware implementation because the expression of the input data forms a binary tree as shown in Fig. 2. The tree in the figure shows a compression mechanism where each node becomes a compressed symbol of the leaves. Although the dictionary size is fixed due to the binary tree and it can be suitable to hardware implementation, the number of nodes in the tree is not deterministic. Thus, we need to invent some additional novel ideas to solve the unpredictable usage of memory to implement streaming hardware-based compression mechanism.

2.3   Discussion

As mentioned above, in order to implement a novel high performance communication data path applying data compression, we need to employ a compression algorithm without (1) data buffering, (2) lookup table exchange and (3) uncertain memory requirement for creating the lookup table. The first condition provides a pipelined data transmission between the compressor and the decompressor. The second one needs a mechanism that counts frequencies of presenting symbol patterns in the input data stream. If a histogram creation can be performed in both sides between the compressor and the decompressor by using the equivalent steps, it is possible to eliminate the table transmission between those. In the machine learning field, we can find well-known algorithm like the lossy counting [9]. The algorithm creates a histogram in a limited number of entries using a linear list data structure, and counts frequency of data patterns. The space saving [11] improves the memory usage of the lossy counting. However, these algorithm uses pointer operations because it is targeted to implement on software. Our approach for the communication data path must be implemented on hardware. Therefore, we need to implement a new method to calculate the histogram. Moreover, the third requirement needs to add a new management mechanism to the histogram creation above based on limited memory resources of hardware because it is not suitable for hardware implementation if the memory usage is dynamically changing. On the other hand, for the data stream of k different symbols, an attractive algorithm for frequency counting has been proposed where the top-θk frequent items are exactly counted within $O(1/\theta)$ space [5] for any constant $0 < \theta < 1$. However, the conventional compression algorithms do not provide any ideal solution for implementing the high speed communication data path, which satisfy the conditions above. In this paper, we propose a novel compression algorithm called LCA-DLT that solves the conditions.

3   Lossless Data Compression Method Based on Hardware

3.1   A Novel Compression/Decompression Algorithm for Streaming Data

We have developed a new compression algorithm called LCA-DLT. It is originally inspired by the LCA algorithm [10] that represents any data patterns using

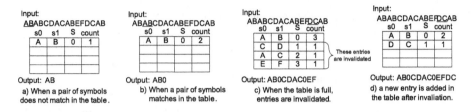

Fig. 3. Compression example of LCA-DLT.

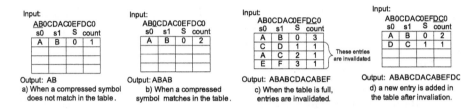

Fig. 4. Decompression example of LCA-DLT.

a binary tree and simplifies the data pattern. In the LCA-DLT, a symbol lookup table is prepared for the compression and the decompression. The table has any number N of entries and the i-th entry E_i includes a pair of the original symbols $(s0_i, s1_i)$, a compressed symbol S_i, and frequent counter $count_i$. The compressor side uses the following rules: (1) reading two symbols $(s0, s1)$ from the input data stream and if the symbols match to $s0_i$ and $s1_i$ in a table entry E_i, after counting up the $count_i$, it outputs S_i as the compressed data, (2) if the symbols do not match to any entry in the table, it outputs $(s0, s1)$ and register an entry $(s0_k, s1_k, S_k, count_k = 1)$ where S_k is the index number of the entry, and (3) if all entries in the table are used, all $count_i$ where $0 \leq i < N$ are decremented until any count(s) become zero and the corresponding entries are deleted from the table.

In the decompressor side, assume that compressed data S is transmitted from the compressor, the steps are the equivalent as the compressor, but the symbol matching is performed based on S_k in an entry. If the compressed symbol S matches to S_k in a table entry, it outputs $(s0_k, s1_k)$. If not, it reads another symbol S' from the compressed data stream and outputs a pair of (S, S') and then the pair is registered in the table. When the table entry is full, the same operation as the compressor side is performed. These operations provide a recoverable histogram on a limited number of lookup table entries.

Let us see an example of compression and decompression operations depicted in Figs. 3 and 4. Assume that the input data stream for the compressor is "ABABCDACABEFDCAB". First, the compressor reads the first two symbols "AB" and tries to match it in the table (Fig. 3(a)). However, it fails matching and registers 'A' and 'B' as the $s0$ and $s1$ in the table. Here, the compressed symbol is assigned in the entry, which is the index of the table "0". Thus, a rule AB→0 is made. The $count$ is initially set to 1. When the compressor continuously reads a

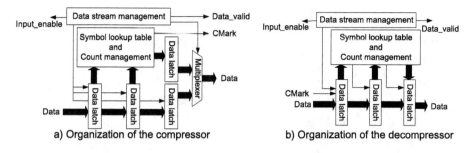

Fig. 5. Organizaitons of the compressor and the decompressor of the LCA-DLT.

pair of symbols (again "AB") and it matches in the table. Figure 3(b) translates "AB" to '0'. Subsequently the equivalent operations are performed. If the table becomes full (Fig. 3(c)), the compressor decrements $count(s)$ of all entries until any $count$s become zero. Here, three entries are invalidated from the table in the figure. The compressor will register a new entry to the invalidated entry from the smallest index of the table. Figure 3(d) shows that the compressor added a new entry after the invalidation. Finally the original input data is compressed to "AB0CDAC0EFDC0".

On the other hand, the decompressor can work as soon as a symbol is outputted from the compressor. The decompressor reads the symbol and tries to match it in the table. In Fig. 4(a), the decompressor reads 'A' first. However, it does not match to any compressed symbol in the table (even though the table is empty). Then it reads another symbol 'B' and registers a new table entry and makes a rule AB→0. The output is "AB" as well as the compressor works. The decompressor reads the next symbol '0' (Fig. 3(b)). It matches to the table entry. The decompressor will translate it to "AB" and output it. After the subsequent decompression operations, when the table becomes full, the decompressor decrements the $count(s)$ as well as the compressor side (Fig. 3(c)). The invalidated entries must equal to the compressor side. Therefore, the compressed symbols are consistently translated to the original symbols. Finally, the inputted compressed data to the decompressor will be translated and outputted as "ABABCDA-CABEFDCAB" that is the same as the input data in the compressor side.

As we can see in the example, the LCA-DLT performs compression/decompression without exchanging the symbol lookup table. Moreover, those operations are performed without buffering in pipeline manner. Thus, the LCA-DLT will work well suitable to the high speed communication data path propagating the compressed data in pipeline manner.

3.2   Hardware Implementation of LCA-DLT

The LCA-DLT can be implemented as shown in Fig. 5. The overall functional block diagrams of the compressor and the decompressor are depicted in Figs. 5(a) and (b) respectively. The input data is propagated via the data latches, and then the compressed and/or the decompressed data is processed in pipeline manner.

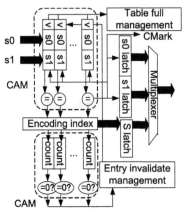
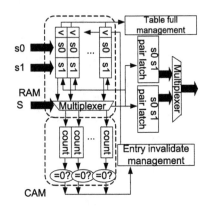

a) Compressor's lookup table and
count management logic

b) Decompressor's lookup table and
count management logic

Fig. 6. Implementation of the managemant parts of the lookup table and the count logic in the LCA-DLT.

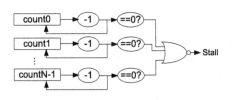
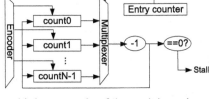

a) An example of the parallel version

b) An example of the serial version

Fig. 7. Decrementing logic for the count(s) to invalidate the table entries(s).

The symbol lookup table performs the compressed/decompressed data translation. The compressor's lookup table receives two input symbols from the latches and outputs the select signal to the multiplexer for the output data. The decompressor's lookup table performs as the opposite data translation.

The lookup table in the compressor is organized as shown in Fig. 6(a). We have designed to obtain 2^n entries in the table when a symbol is n bits. The matching part for $s0$ and $s1$ must be organized as a CAM (Content Addressable Memory), which outputs the index (i.e. the address in the CAM) matched to an inputted pair of ($s0$, $s1$). Here, the index equals to the compressed symbol. The management part for $count$ is also organized by a CAM. The enable signal from the matching part counts up the corresponding $count$. The full management logic of the lookup table activates the invalidate control that decrements the $count$. Then it resets the valid bits (v in the figure) corresponding to the invalidated entry.

The lookup table in the decompressor is organized with a RAM and a CAM as depicted in Fig. 6(b). The management part of $count$ also performs as equal

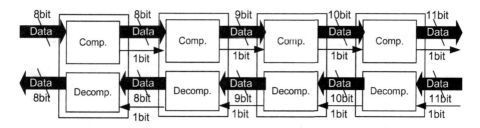

Fig. 8. Cascading compressor/decompressor module

as the compressor's implemented based on a CAM. However, the matching part is implemented simply by a RAM in which the compressed data as the address is inputted and the original uncompressed data pair is outputted.

The invalidate operation to find the minimum *counts* in the table entries by decrementing it can be implemented in two options. Both implementations output the stall signal to stop the compression/decompression data pipeline. To achieve good performance of compression/decompression, it is important to reduce the number of clock cycles when the stall signal is active. The first one is the implementation based on parallel decrement logic and full comparison with zero as illustrated in Fig. 7(a). Although this implementation requires a large resource, it can reduce the required clock cycles for the invalidation operation due to the parallel evaluation of the counts. Another is the one based on serial decrement logic with rounding the indices of the entries for one by one at every clock as shown in Fig. 7(b). This needs a set of a decrementing logic and a comparator with zero. However, it needs more clocks to find invalidation candidates. These two implementations are tradeoffs between the amount of logics and the compression speed when the table becomes full.

When we support 8 bit data input for the compressor, it must support any 8 bit data to compress and translate it to compressed symbols. Here we need to pay attention to the compressed symbol that must be unused in the input data stream. However, it is impossible to define the unused symbols because any 8 bit data can be inputted to the compressor. Therefore, the compressor adds the *compression mark bit* (CMark) that indicates if the symbol is compressed or not.

Moreover, combining the compressor and the decompressor in a module and cascading the modules as illustrated in Fig. 8, we can compress long symbol patterns corresponding to 2, 4, 8, 16 symbols respectively when the number of modules is four. If the input data at the first compressor is 8 bit long, the output compressed data becomes 12 bits after four modules due to the CMark bits.

3.3   Communication Protocol for LCA-DLT

The LCA-DLT must begin to compress/decompress symbols using an initialized table on both sides because it must synchronize the dynamic histogram creation. When we apply it to a communication data path, there are two types of system organization. One has a shared reset line in a system. A typical system is

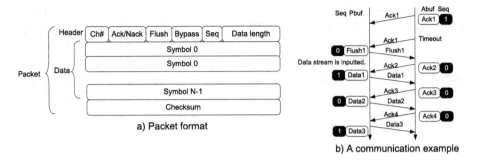

Fig. 9. Communication protocol for LCA-DLT.

a memory controller on a motherboard. Before sending a data stream via the LCA-DLT, the system can synchronize the initial content of the lookup table just by asserting the reset line shared by the compressor/decompressor. Another is a system where the compressor and the decompressor are placed remotely such as connecting those by a copper wire via a communication physical layer. The system needs a communication protocol to synchronize the lookup table.

The protocol works as a packetization of data stream for the compression and the decompression. The packet is illustrated in Fig. 9(a). The compressor acts as the master and sends data/commands to the decompressor. The header has a channel number, an ack/nack, a flush, a bypass, a sequence and a data length. The packet has data and the checksum. The channel number specifies one of channels for communication data paths. The flush bit commands to initialize the tables. The bypass bit indicates the packet includes uncompressed data. The sequence bit sends the sequence register state of the packet sender. Finally the data length includes the number of symbols in the packet.

The compressor side has *Pbuf* that saves the previous sending packet, *Seq* register that toggles the state. The decompressor side has *Abuf* that saves the previous Ack/Nack, the *Seq* register and a timeout counter. Figure 9(b) shows an example flow of the protocol. After reset, the compressor/decompressor resets/sets the Seq register. The decompressor sends an Ack packet repeatedly after timeout. First the compressor sends a Flush packet to initialize the table. When the decompressor receives the *Flush* packet correctly, it initializes the table, sends an Ack packet and toggles the Seq register. When the compressor side receives the *Ack* packet, it toggles the Seq register. When any data is inputted to the compressor, it performs the compression, makes a *Data* packet and sends it. The subsequent packet exchange is the same as the Flush mechanism toggling the Seq register. If the receiving Seq bit differs from the Seq register, the packet is dropped. This promotes the timeout in the decompressor and thus the retransmission is activated.

As proposed in this section, the LCA-DLT helps to implement the dynamic histogram generation and provides the compression/decompression mechanism without sending the symbol lookup table. Therefore, it does not require any data buffering memory resource. Thus it is expected to implement a high speed

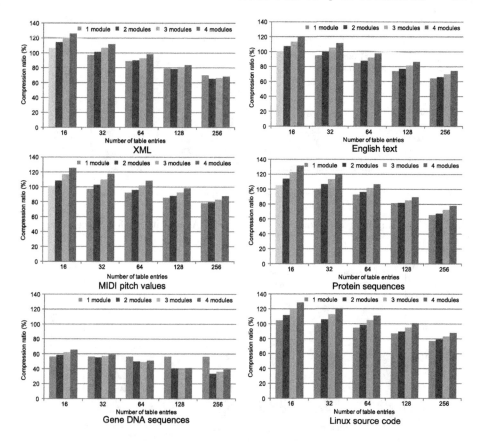

Fig. 10. Compression performance of LCA-DLT.

communication data path based on hardware that was not able to realize the
stream-based compression mechanism applying the conventional ones.

4   Performance Evaluation

Let us evaluate the static and dynamic performances of LCA-DLT. The static
one will show the compression ratios using a software emulator. The dynamic
one shows the hardware performances. Although the LCA-DLT treats continuous data stream, in order to evaluate a quantitative situation, we used several
patterns of 10 Mbyte benchmark data (the number of clock cycles for transfer at eight bits without compression takes 10240 cycles) of the text collection
listed in [1]: Linux source code, MIDI pitch values, protein sequences, gene DNA
sequences, English texts and XML.

4.1   Performance of Compression Algorithm

Figure 10 shows the compression ratios ($(compressed\_data\_size/original\_$
$data\_size) \times 100$) when the numbers of the table entries and the ones of the

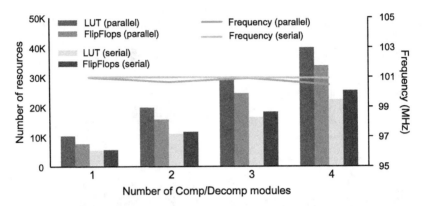

Fig. 11. The numbers of resources used on FPGA and its performance based on frequency.

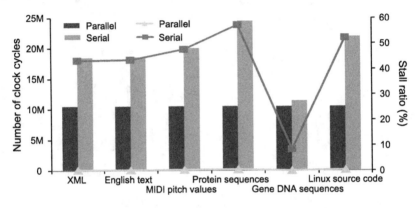

Fig. 12. The Numbers of clock cycles to compress a 10 MByte data and ratios of its stall cycles.

compressor and the decompressor modules linearly cascaded as depicted in Fig. 8. The numbers of the table entries are changed from 16 to 256. Focusing on the performance impact of the number of the table entries, the compression ratios are improved linearly except the gene DNA sequence. Because the DNA data has a few patterns, all patterns can be saved in 16 entries. On the other hand, focusing on the impact of the number of the modules, the compression ratios degraded in the case of more than two modules. This means that a communication data path using too many compression modules becomes disadvantage due to the CMark bit added after each module.

As we can see in the graphs, the LCA-DLT is able to adjust the compression ratio changing the number of table entries and the number of compression modules. These provide large contributions of which the hardware designer can select the best combination under the restriction of limited hardware resources regarding memory bits and logics.

4.2   Performance of Hardware Implementation

Figures 11 and 12 show the hardware performance of the LCA-DLT. We compiled the hardware logic of 8 bit data input at the first compressor of the cascaded modules for the Xilinx Artix7 device (XC7A200T-1FBG676C). Figure 11 depicts a comparison between the parallel and the serial invalidation mechanisms regarding the maximum frequencies and the resource usage when the number of modules were varied from one to four. Here, the graph does not include the communication protocol part. It was implemented by only hundreds of slices and a memory block in the FPGA. The LCA-DLT works at about 100 MHz with any number of the modules. This means that the compressor and the decompressor achieve 800 Mbit/sec interface where the data communication path can be implemented by the lower bandwidth if the data is compressed. On the other hand, the LCA-DLT gives a large impact for resource usage regarding the logic but the memory because the recent FPGA does not have any dedicated hardware macros for CAMs. It is inevitably implemented by LUT and registers in the FPGA. We also compared the hardware sizes of the parallel/serial invalidation mechanisms. The parallel version uses larger hardware resources as we expected.

Regarding the dynamic performances of the LCA-DLT, Fig. 12 shows the performance comparison between the parallel and the serial invalidation mechanisms with two modules. We focused on the numbers of clock cycles consumed by compressing data stream and the stall ratios during the entry invalidation mechanism. These are counted at the input of the first compressor of the cascaded modules. During the stall, any data is not inputted to the compressor and all the compression/decompression processes stop. The parallel version includes very few stall. However the hardware resource explodes. Assume that the hardware works at 100 MHz, the effective bandwidth in the input of the first compressor is about 800 Mbit/sec and 340–730 Mbit/sec in the case of the parallel and the serial invalidation respectively. If we consider an ASIC implementation of LCA-DLT, we are expecting to achieve 20–30 times faster implementation than the FPGA [7]. It will achieve a communication data path of 16–24 Gbit/sec at 2 GHz to 3 GHz working clock. When we consider the compression ratio shown in Fig. 10, the output bandwidth of the second compressor will be reduced to 35–80 % of the original data size. This means that the LCA-DLT realizes a communication data path that can send more data even if the speed of the path is slow, and also contributes largely to realize a high speed communication data path using data compression mechanism providing flexible adjustment between the hardware resources and the compression performance.

5   Conclusions

This paper described an implementation of communication data path by applying the data compression mechanism based on hardware implementation. We have proposed a novel compression algorithm called LCA-DLT. It provides a stream-based compression by managing a dynamic histogram. The LCA-DLT provides a mechanism that best fits to hardware-based implementation. The performance evaluations have shown the flexibilities for the hardware designer between the resources and the performance. For the future works, we are planning to apply the compression mechanism to the communication data path of applications on high performance computing. It would provide a faster computing performance in a slower communication physical medium.

Acknowledgment. This work is partially supported by JSPS KAKENHI Grant Number 15H02674, 26280088 and JST CREST.

References

1. http://pizzachili.dcc.uchile.cl/
2. Apostolico, A., Lonardi, S.: Off-line compression by greedy textual substitution. Proc. IEEE **88**(11), 1733–1744 (2000)
3. Grossi, R., Gupta, A., Vitter, J.S.: High-order Entropy-compressed Text Indexes. In: Proceedings of the Fourteenth Annual ACM-SIAM Symposium on Discrete Algorithms / SODA 2003, pp. 841–850. ACM (2003)
4. Jacobson, G.: Space-efficient static trees and graphs. In: Proceedings of 30th IEEE Annual Symposium on Foundations of Computer Science, pp. 549–554. IEEE (1989)
5. Karp, R.M., Shenker, S., Papadimitriou, C.H.: A simple algorithm for finding frequent elements in streams and bags. ACM Trans. Database Syst. **28**(1), 51–55 (2003)
6. Kieffer, J., Hui Yang, E.: Grammar-based codes: a new class of universallossless source codes. IEEE Trans. Inf. Theor. **46**(3), 737–754 (2000)
7. Kuon, I., Rose, J.: Measuring the gap between FPGAs and ASICs. IEEE Trans. Comput. Aided Des. Integr. Circ. Syst. **26**(2), 203–215 (2007)
8. Larsson, N., Moffat, A.: Offline dictionary-based compression. In: Proceedings of Data Compression Conference (DCC 1999), pp. 296–305. IEEE, March 1999
9. Manku, G.S., Motwani, R.: Approximate frequency counts over data streams. In: Proceedings of the 28th International Conference on Very Large Data Bases, pp. 346–357. VLDB Endowment (2002)
10. Maruyama, S., Sakamoto, H., Takeda, M.: An online algorithm for lightweight grammar-based compression. Algorithms **5**(2), 214–235 (2012)
11. Metwally, A., Agrawal, D., Abbadi, A.E.: An integrated efficient solution for computing frequent and top-k elements in data streams. ACM Trans. Database Syst. **31**(3), 1095–1133 (2006)
12. Nevill-Manning, C.G., Witten, I.H.: Identifying hierarchical structure in sequences: A linear-time algorithm. J. Artif. Intell. Res. **7**(1), 67–82 (1997)
13. Vitter, J.S.: Design and analysis of dynamic huffman codes. J. ACM **34**(4), 825–845 (1987)
14. Yamagiwa, S., Aoki, K., Wada, K.: Performance enhancement of inter-cluster communication with software-based data compression in link layer. Proc. IASTED PDCS **2005**, 325–332 (2005)
15. Ziv, J., Lempel, A.: A universal algorithm for sequential data compression. IEEE Trans. Inf. Theor. **23**(3), 337–343 (1977)
16. Ziv, J., Lempel, A.: Compression of individual sequences via variable-rate coding. IEEE Trans. Inf. Theor. **24**(5), 530–536 (1978)

Author Index

Printed in the United States
By Bookmasters